AF430920

ACHARYA N.G. RANGA AGRICULTURAL UNIVERSITY
DEPARTMENT OF FARM FORESTRY

PRACTICAL MANUAL ON
PRINCIPLES AND PRACTICES OF SOCIAL FORESTRY

(COURSE NO. AGRO : 302)

DEPARTMENT OF FARM FORESTRY
COLLEGE OF AGRICULTURE, RAJENDRANAGAR
HYDERABAD - 500 030

Practical Manual on
Principles and Practices of Social Forestry

Dr. B. Joseph
Professor & Head,
Department of Farm Forestry,
College of Agriculture, Rajendranagar,
ANGRAU, Hyderabad.

Dr. S. Hemalatha
Associate Professor,
Department of Farm Forestry,
College of Agriculture, Rajendranagar,
ANGRAU, Hyderabad.

Dr. A. Madhavi Lata
Assistant Professor,
Department of Farm Forestry,
College of Agriculture, Rajendranagar,
ANGRAU, Hyderabad.

Dr. P. Laxminarayana
Professor,
Department of Farm Forestry,
College of Agriculture, Rajendranagar,
ANGRAU, Hyderabad,
A.P., India.

BSP **BS Publications**
A unit of **BSP Books Pvt. Ltd.**
4-4-309/316, Giriraj Lane, Sultan Bazar,
Hyderabad - 500 095 - A.P.
Phone : 040 - 23445605, 23445688

Practical Manual on Principles and Practices of Social Forestry

© 2014 *by Publisher*

Published by

 **BS Publications**
A unit of **BSP Books Pvt. Ltd.**
4-4-309/316, Giriraj Lane, Sultan Bazar,
Hyderabad - 500 095 - A.P.
Phone : 040 - 23445605, 23445688
e-mail : info@bspbooks.net
www.bspbooks.net

ISBN: 978-93-90211-05-0

Dedicated to

My Parents

Late Sri B. Subba Rao and Smt. B. Rattalu

– B. Joseph

ACHARYA N.G. RANGA AGRICULTURAL UNIVERSITY

DEPARTMENT OF FARM FORESTRY

<u>CERTIFICATE</u>

This is to certify that this is a bonafied record of practical work done by

Mr. / Ms. ________________________ I.D. No. _______________during

the First / Second semester of the academic year _______________.

Date :   Course-in-charge

ACHARYA N.G. RANGA AGRICULTURAL UNIVERSITY

Administrative Office, Rajendranagar,
Hyderabad – 500 030. A.P. INDIA
Phone: (O) 040-24015197 (O & F)
Mobile: + 91-9848346669
Email: prof.tvksingh@yahoo.com
deanagri@hotmail.com
Grams: "AGRIVARSITY"

FOREWORD

Current awareness on climate change and global environment made us realize the importance of forests and the dire need for its protection and management. Forests are continued to be exploited indiscriminately for fuel wood, fodder and timber etc., across the world several countries have realized consequences of potential harmful influence of widespread deforestation, excessive grazing lead to continued ecological imbalance.

For maintaining and enhancing productivity of forests and to increase forest cover in non-forest land, sound knowledge of "Social Forestry" is a must. Research work on agroforestry and social forestry in India aimed at protecting global environment resulted in bringing out strategies to be adopted in ameliorating harmful effects due to deforestation and urbanization as well as increase the ability of tree species to the changes in environment.

Dr B. Joseph made an attempt in bringing out a manual on **"Principles and Practices of Social Forestry"** to impart knowledge to the undergraduate students on the basic and applied aspect of social forestry who have just entered the field of Agriculture. I hope this manual will be of immense use to the beginners as well as fellow teachers in understanding the subject with hands-on experience.

T.V.K. SINGH

Place - Hyderabad
Date - 15.04.2014

Preface

Forests are one of the natural resources needed by human beings for daily needs. They are crucial in preserving earth's ecological balance and hence the global concern on environmental protection. With the ever increasing human population there has been a tremendous pressure on forests as forests lands are cleared for agriculture, urbanization, military purpose, power projects, wood purpose, fire wood requirements. This had lead to land and forests degradation including threat to wildlife. Taking these factors into consideration Governments world wide implemented social forestry. As per the "National Forest Policy 1988, India aims at maintaining $1/3^{rd}$ of its total land area under forest with a view to protect forests and ensure sustained supply of raw material." Social forestry, defined as "of the people, by the people and for the people", means the management and protection of forests and afforestation. The National Commission on Agriculture under the concept of social forestry emphasised an operational model ensuring the continuous supply of fuel, fodder and small timber to the local population meeting their daily needs of the local people residing in rural areas and nearby forest areas and to combat the environmental problems.

Knowledge on the scope mission, classification of tree species, their economic and environmental benefits of social forestry are taught in detail in the under graduate course "Principles and Practices of Social Forestry" as per the ICAR V Deans committee Recommendations from 2007. The present manual on the above subject is means for students describes in detail the identification of tree species, handling and pre-sowing treatments of forest tree seeds, plantation and aftercare, biomass yield of trees and about major and minor products etc along with glossary. It is hoped that the practical experience gained from the study of this manual will be of immense use for agricultural students for learning social forestry. I take this opportunity to gratefully acknowledge my senior Colleague Dr M.V.R. Subramanyam in gathering the valuable information need to prepare this manual and I also thank the Dean of Agriculture, ANGRAU, Hyderabad for the guidance and encouragement in preparation of this manual. I also acknowledge and express my sincere thanks to the ICAR for providing necessary financial assistance from ICAR funds of the college of Agriculture, Rajendranagar, Hyderabad. Last but not least I thank all my post graduate students Ms. CH. Pallavi and P. Arun Babu who have taken pains in arranging the manual in a systematic manner and made ready for publishing.

I also thank my wife Mrs. B. Radha for continuously supporting me in bringing this manual for students usage.

Date: 14/4/2014.

- B. JOSEPH

Contents

Foreword...(ix)

Preface...(xi)

Ex. No.	Title	Page No.
1	Identification of Tree Species Suitable for Timber, Fuelwood and Fodder	1
2	Identification of Tree Species Suitable for Roadside Plantation, Field Bunds, Windbreaks and for Wastelands	25
3	Identification of Minor Forest Tree Species, Trees for Beautification Purpose and Nitrogen Fixing Tree Species and Other Species Suitable for Agroforestry	45
4	Identification of Seeds of Important Tree Species	53
5	Collection, Extraction and Storage of Tree Seeds	61
6	Tests for Seed Viability and Germination	81
7	Application of Pre-sowing Seed Treatment to Tree Seeds	93
8	Preparation of Nursery Beds and Sowing	107
9	Transplanting, Raising of Bare-Rooted and Containerized Plants	123
10	Field Planting Techniques and Aftercare (Tending)	137
11	Biomass Estimation in Energy Plantations and Tree Height Measurement	157
12	Field Evaluation of Different Agroforestry Systems	167
13	Field Evaluation of Shelterbelts & Windbreaks	173
14	Major Forest Products	179
15	Cost of Cultivation of Commercial Trees in Wastelands: Bamboo	193
16	Calculations	197
	Glossary	203

Index

S. NO	NAME OF THE EXERCISE	DATE OF CONDUCTING EXERCISE	DATE OF SUBMISSION	SIGN. OF COURSE I/C

S. NO	NAME OF THE EXERCISE	DATE OF CONDUCTING EXERCISE	DATE OF SUBMISSION	SIGN. OF COURSE I/C

IDENTIFICATION OF TREE SPECIES SUITABLE FOR TIMBER, FUELWOOD AND FODDER

Objectives:

1. ____

2. ____

3. ____

(A) Timber wood species

1. Teak (*Tectona grandis*)

Vernacular names: Hindi – Sagun or Sagwan, Telugu – Teku.

Family: Verbinaceae

Teak is a very valuable tree and it is called as king of timbers. In Greek Tectona means "Carpenter" and grandis means "Large size". It is a deciduous tree that grows in all parts of the country.

Morphological description: The branches are four sided at the young age and covered with hairs. Young trees have a large sized (40-60 cm long and 20-30 cm broad) leaves like hand winnowers. Leaves are simple rough, ovate and pointed at the tip. Flowers are small white colored but emerge in large, open clusters from the tips of branches. The fruit is hard bony nut enclosed in thick sponges. Seeds are very small and have a strong dormancy.

Leaf

Flower

Fruit

Economic importance: It is extensively used for house building, bridge constructions and for making poles, beams, roofs, doors, window frames, railway sleepers etc., and also for furniture. Leaves are used for thatching houses in tribal areas.

2. Rose wood (*Dalbergia latifolia*)

Vernacular names: Hindi – Shisham, Telugu – Jitregi, English – Bombay black wood.

Family: Leguminosae.

This elegant tree is deciduous and a warm loving and grown luxuriously in moist soils. Rose wood grows on a wide range of soils from gravel to loamy soils. But thrives best in well-drained, deep, moist loamy soils.

Morphological description: Leaf contains 3-7 leaflets and they are elliptical obtuse, sometimes emerginate. The flowers are white and very small, axillary. Fruit is flattish, oblong pod and also lancelolate. Immature pods are light green and mature are brown in colour and indehiscent.

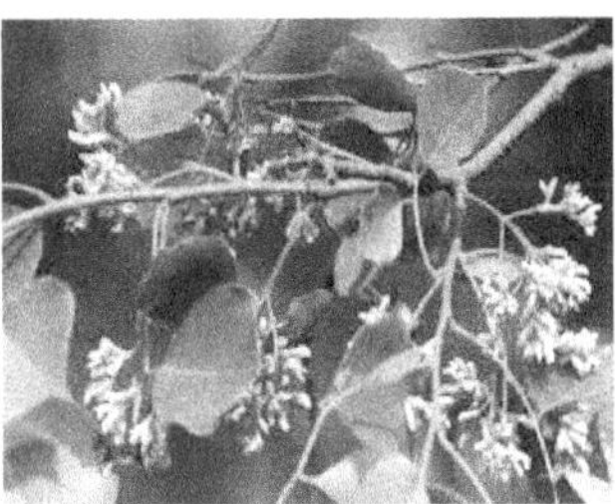

| **Leaf** | **Fruit/pods** | **Flowers** |

Economic importance:

The wood is very heavy and stronger than that of teak. It is used for cabinet and furniture making. Rose wood is a valuable timber for carving, veneers and plywood boards. Leaves are nutritious and a good fodder for cattle and goats.

3. Sissoo (*Dalbergia sissoo*)

Vernacular names: Hindi – Shisham, Telugu – Errasissu.

Family: Leguminosae

The sissoo is a native of tropical India. It is also a large deciduous tree and grown in wide range of soils.

Morphological description: The leaf contains 3-5 leaflets and leaves are broadly elliptic or ovate and acuminate. The flowers are yellowish, white, and axillary nearly sessile. Fruit is a pod which is flattish linear and lanceolate. Immature pods are light green and mature pods are brown and indehiscent.

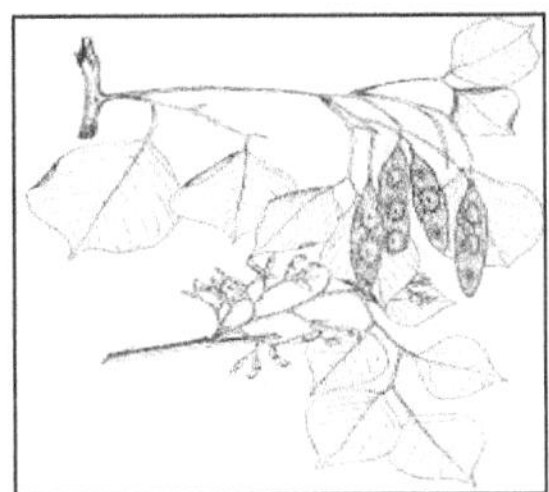

| **Leaf** | **Twig** | **Flower** | **Pods** |

Economic importance: The timber is heavy and used for high-class furniture. It is extensively used for cabinet wood, railway sleepers, musical instruments, lorry bodies etc. and also used for house construction. The wood gives good quality paper pulp. The green leaves and pods can also be used as cattle feed and the dried branches are considered as excellent firewood.

4. Sal *(Shorea robusta)*

Vernacular names: Hindi – Sal, Telugu – Guggilam or gugal.

Family: Dipterocarpaceae

The tree is large and mostly found in deciduous forests. Sal grows on a wide range of soils but attains maximum size on deep, well drained and moist soils. Both the hilly tracts and plains are equally suitable for its growth.

Morphological description: Leaves are glabrous when fully grown and broad – ovate, more or less acuminate ending **in a obtuse** point. Flowers are yellowish on short pedicels arranged in large compound axillary and terminal panicles. Fruits are obtuse, oblong or spathulate.

   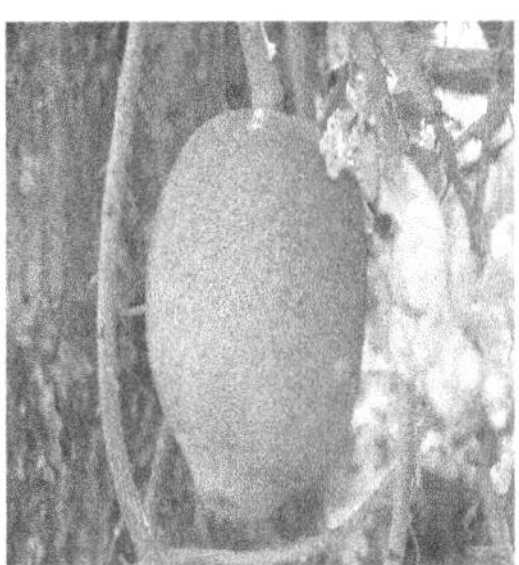

Economic importance: The aged timber is exceedingly strong and durable and mostly used as railway sleepers and rail carriages. The poles of sal are used for overhead electric and telegraphic lines. The wood is used for beams, cart wheels etc. The bark and leaves are used for tanning.

5. Sain *(Terminalia tomentosa)*

Vernacular names: Hindi – Sain, Telugu – Nallamaddi.

Family: Combretaceae.

It is a large, evergreen tree and has a broad crown and trunk tall with rough bark. It is grown in wide range of soils.

Morphological description: Leaves are hard, elliptic or ovate arranged opposite. The upper most leaves often alternate. 1 or 2 glands appear on back of the leaf near base of the midrib and under side is soft tomentose rarely glabrous. Branchlets are covered with short rust coloured pubescence. Flowers are dull yellowish colour in erect terminal panicles. Fruit is woody, winged with 5 brown wings.

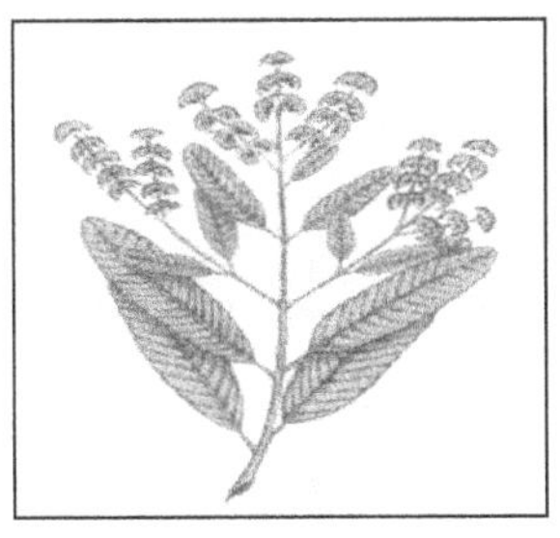

| Branch with flowers | Leaf | Fruit |

Economic importance: It is strong and moderately heavy wood. Wood is used for house construction, beams, agricultural implements, boat building etc. Bark is used for tanning the hides.

6. Arjun (*Terminalia arjuna*)

Vernacular names: Hindi – Arjuna, Telugu – Tellamaddi or yerramaddi.

Family: Combretaceae

Arjun is a beautiful and evergreen tree. The trunk is buttressed. The tree has a broad crown and branchlets are droopy with smooth bark. The tree is found is tropical moist deciduous and dry deciduous forests of India.

Morphological description: Leaves are thick and oblong or obovate. Flowers are dull yellowish colour in erect terminal panicles. Branchlets are droopy with smooth bark. Fruit is long woody structure with wings.

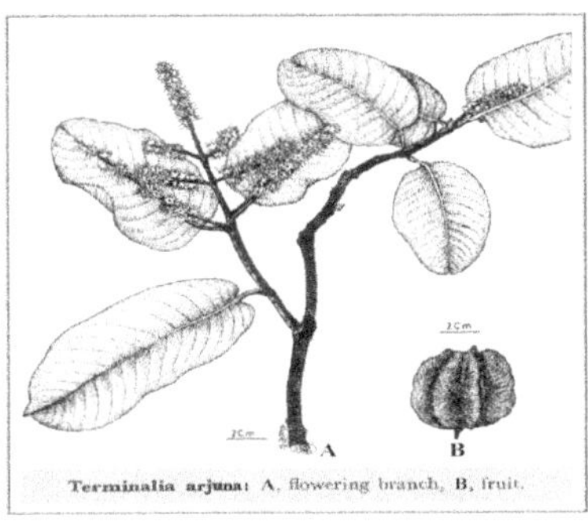

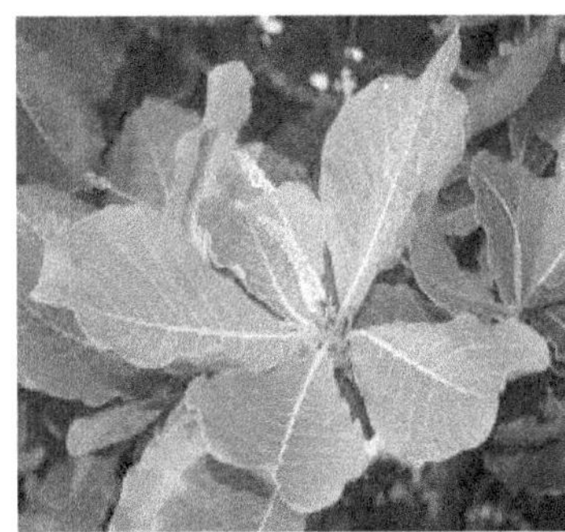

Economic importance: The wood is strong and moderately heavy. Wood is used for making agricultural implements, boat building, carts etc. and suitable for plywood manufacture. It is also used for building houses, electric poles etc. Bark is used for tanning the hides. The foliage is also a good fodder for cattle.

7. Silver Oak (*Grevillea robusta*)

Family: Proteaceae

It is a large tree with young shoots, rusty tomentose.

Morphological description: Leaves are pinnate and long deeply pinnatified so that the leaves are almost bipinnate and sometimes tri-pinnate. Flowers are orange coloured axillary. Fruit is an oblique coriaceous and dehisive.

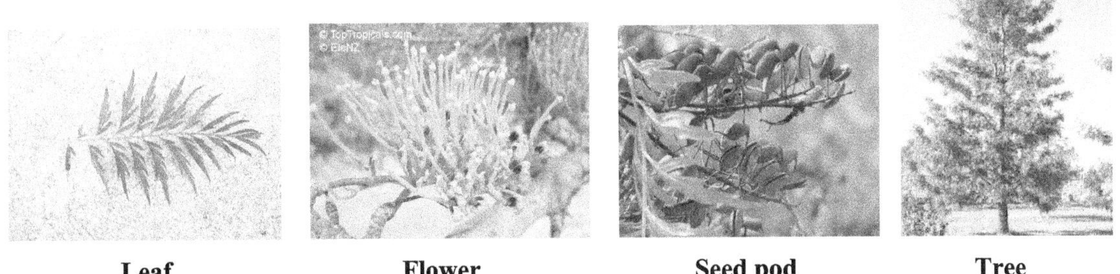

| Leaf | Flower | Seed pod | Tree |

Economic importance: Wood is used for small timber and manufacture of packing boxes and matchsticks. It is used as shade trees for tea and coffee plantations.

8. Axle wood *(Anogeissus latifolia)*

Vernacular names: Hindi – Dhawa, Telugu – yellamaddi or chirimanu.

Family: Combretaceae

Axle wood is very common in the Himalayan foothills. It is moderate to large sized deciduous tree. The trees are also found in deciduous forests of Madhya Pradesh, Maharashtra, Tamilnadu and Andhra Pradesh.

Morphological description: Leaves are elliptic, ovate or obovate, lanceolate and arranged in alternate manner. Branches are drooping and flowers are small with globose head, borne in short axillary racemes. Fruit is a dry drupe, compressed, 2 winged.

Leaf and Flowers

Economic importance: Wood is used for making poles, house posts, plough handles etc. It is also used extensively for construction and furniture making.

9. Red sanders *(Pterocarpus santalinus)*

Vernacular names: Hindi – Lalchandan, Telugu – Yerrachandanam and rakthagandam.

Family: Santalaceae

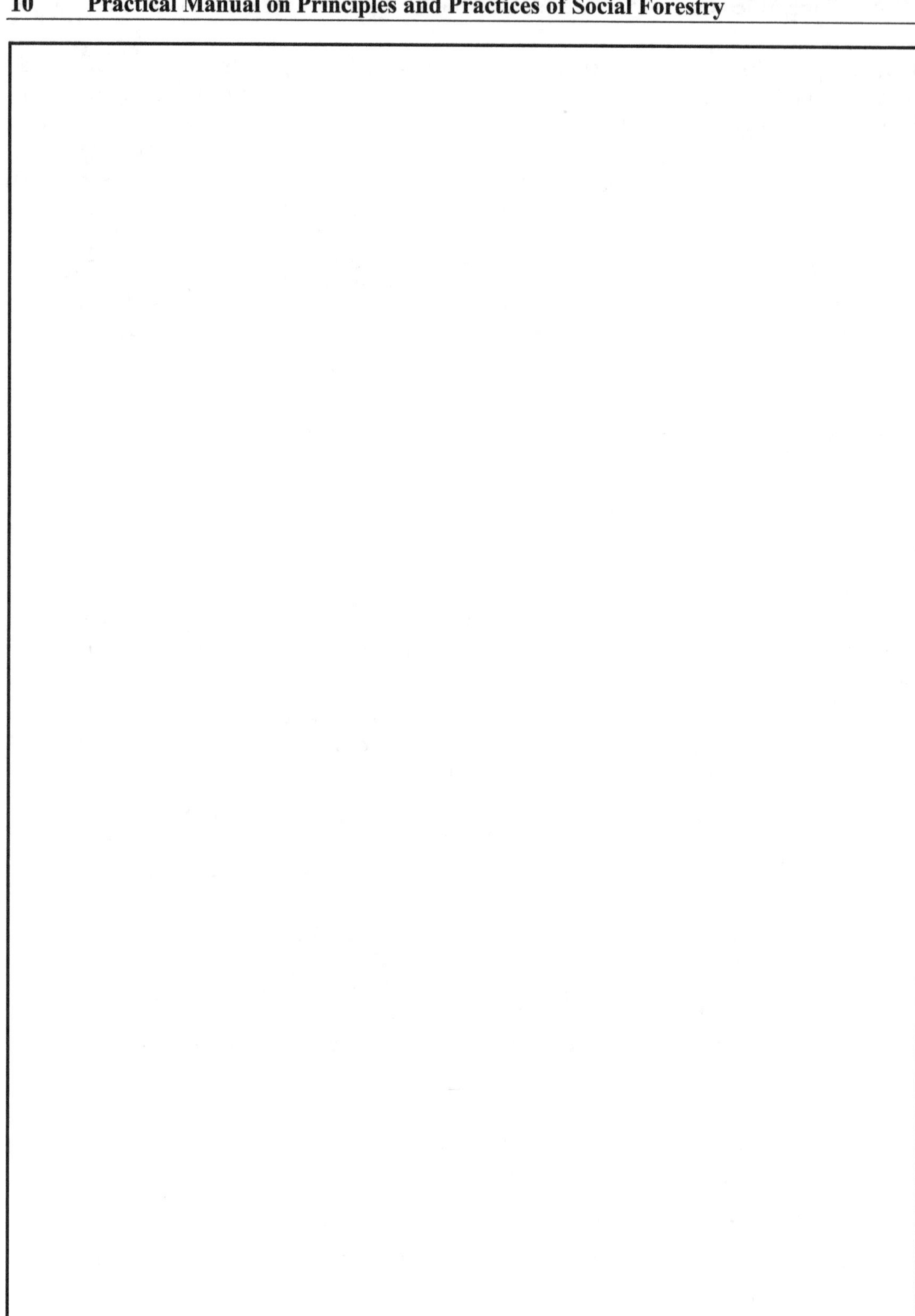

Red sanders are the most valuable species growing in dry deciduous forests of peninsular India. The tree attains moderate height and well branched.

Morphological description: Leaves are compound and leaflets are nearly circular in shape. Flowers are yellow and were in bundles. Fruits are flat with a slight curve.

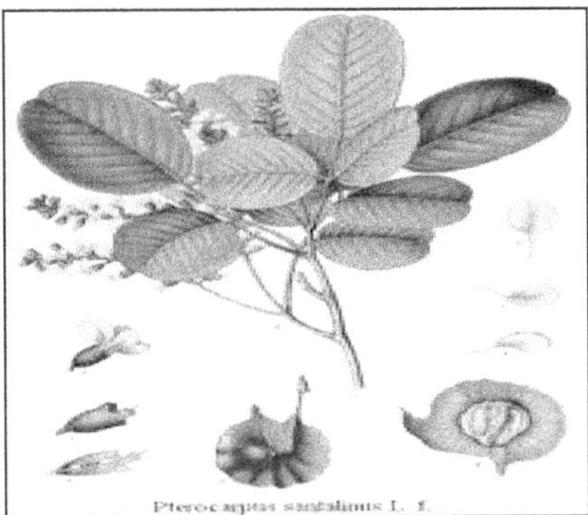

Branch with leaf, flowers and pod

Economic importance: The wood is very hardy and highly priced for house posts, poles, picture frames, boxes, agricultural implements etc. Heartwood is used for making musical instruments.

10.   Gumhar *(Gmelina arborea)*

Vernacular names: Hindi – Gumhar, Sewan, Telugu – Gumarteku, Gummadi, Teku.

Family: Verbinaceae

Gmelina arborea is distributed generally throughout India except the very dry regions of Rajasthan and Gujarat. It is fast growing tree.

Morphological characters: Leaves fall in January and February and are large in size, acuminate. The panicles of flowers appear in February. Fruits are succulent ovoid or oblong drupe with leathery shining pericarp, sweetish pulp and hard long stone.

Leaf	Flower	Fruit

Economic importance: It is a good timber tree species tind hardy. The leaves are also used as good fodder.

(B)   Fuelwood species

1. Siris (*Albizzia lebbek*)

Vernacular names: Hindi – Siris, Telugu – Dirisanam, English – East Indian Walnut.

Family: Leguminosae

Siris is a medium to large sized deciduous tree.

Morphological description: Leaves divided into several leaflets and are obliquely oblong, obtuse and reticulate. Flowers are greenish white colour with jasmine like fragrance, terminal. Fruit is a flattish thin pod, straw coloured when matured and dehiscent. Fruit remain conspicuous to the tree even after leaf fall.

| **Leaf** | **Branch** | **Flower** |

Economic importance: The wood is hard and durable and used for making agricultural implements, carts and also furniture.

2. Kassod *(Cassia siamea)*

Vernacular names: Marathi: Kassod, Telugu – Sematangedu, English – Yellow cassia.

Family: Leguminosae

It is middle sized, sometimes grown as large tree. It is not browsed by the animals and suitable for wasteland afforestation. It is a native of South East Asia and was introduced in India as an avenue tree.

Morphological description: Leaves are in pairs of leaflets, elliptic, oblong, flowers are yellow coloured, arranged in large terminal panicle, racemes often **corymbose.** Fruit is cylindrical pod indehiscent, clusters hanging at the end of branches.

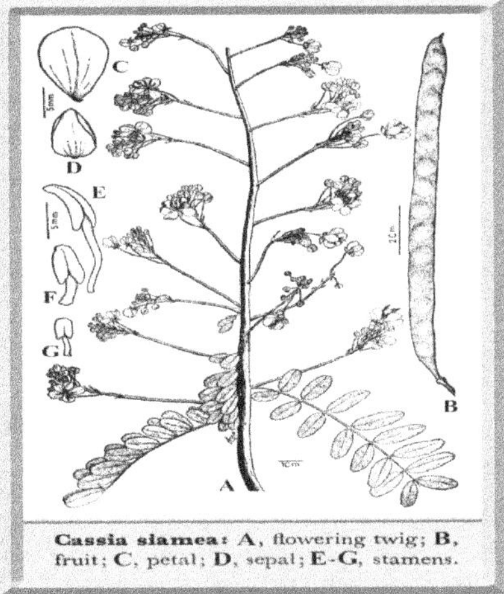

Cassia siamea: A, flowering twig; B, fruit; C, petal; D, sepal; E-G, stamens.

Economic importance: Wood makes excellent fuelwood. The heartwood is used as timber for cabinet making and also used for shelves, walking sticks etc.

3. Akashmoni *(Acacia auriculiformis)*

Family: Leguminosae (Mimosoideae)

It is a small medium sized evergreen tree. This species is a native of North Australia and was introduced in India as avenue tree.

Morphological description: Leaves are phyllode type and leaf stock is modified into a flattened blade. Flowers are white to yellow coloured, tiny and fragrant terminal. Fruit is a wide pod, flat, hard and almost woody.

Flower

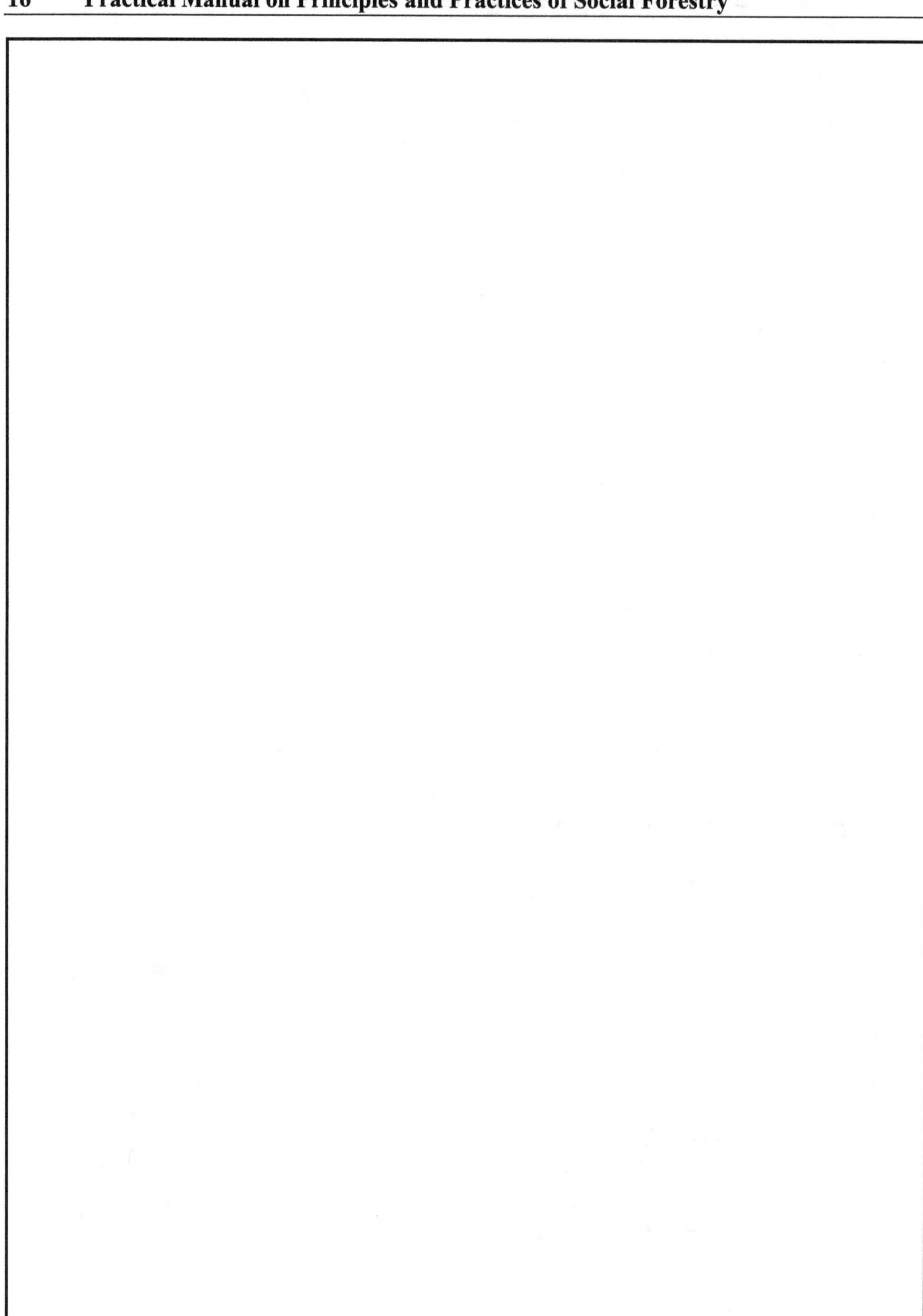

Economic importance: The wood is well suited for fuelwood. This tree is widely planted as an ornamental and shade tree owing to its dense evergreen foliage.

4. Babul *(Acacia nilotica. Syn: Acacia arabica)*

Vernacular names: Hindi – Babul, Telugu – Nalla tumma.

Family: Leguminosae (Mimosoideae)

It is a moderate sized, spiny and evergreen tree. It is indigenous to India and Pakistan.

Morphological description: Leaves are pinnate and leaflets are small. Flowers are golden yellow coloured in globose heads in axils and fruit is a solitary and monoliform pod with whitish tomentose, indehiscent.

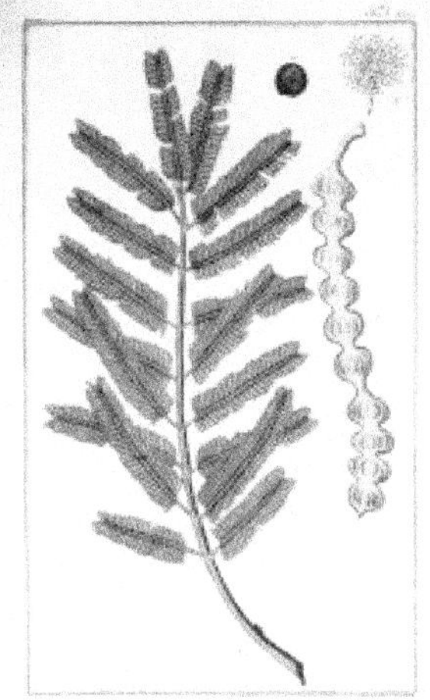

Twig with flowers   **Pods**

Economic importance: It is used as fuelwood and makes excellent quality of charcoal. It is also used as small timber for agricultural implements etc.

5. Mesquite bean *(Prosopis juliflora)*

Vernacular names: Hindi – Vilayati kikar, Telugu – Sarcar tumma.

Family: Leguminosae (Mimosoideae)

It is a thorny deciduous tree or bush sometimes semi-evergreen. This species is a native of Central America and Northern South America distributed in arid parts of the world, widely propagated in India.

Morphological description: Leaves are dark green, bipinnated and leaflets opposite. Flowers are creamy white in colour in axiles and sweet fragrant. Fruit is a succulent pod and seeds are flattened.

| Leaf | Flower | Pods |

Economic importance: It is used as fuelwood and makes super quality charcoal. It is also used for fence posts. Pods are used for cattle feed.

(C)   Fodder species

1. Subabul : *(Leucaena leucocephala)*

Vernacular names: Hindi – Subabul, Telugu – Khari, English – Horse Tamarind.

Family: Leguminaseae (Mimosoideae)

It is a large tree and evergreen and it is very hardy species grown in wide range of soils. There are 13 species of the genus *Leucaena* and the most common species is *Leucaena leucocephala.* These are native trees of America.

Morphological description: Leaves are feathery, pinnate with 10 -15 pairs of leaflets. Flowers are small, white in dense globose heads. Fruit is a flat pod appearing in bunches. Mature pods are brown and shining.

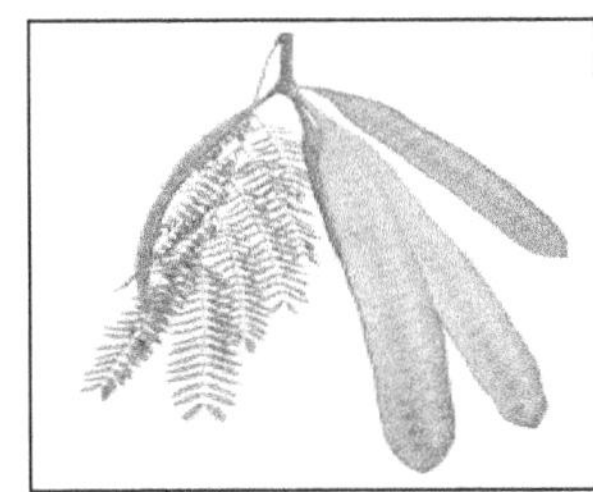

| Twig | Flower | Pods |

Economic importance: Leaves, pods and seeds are nutritious and digestible and relished by cattle, sheep and goat. Wood is medium hard and it works well for carpentry. It is also used as fuelwood.

2. Sesbania *(Sesbania grandiflora)*

Vernacular names: Hindi – Agasti, Telugu – Avisa.

Family: Leguminosae (Papilionoideae)

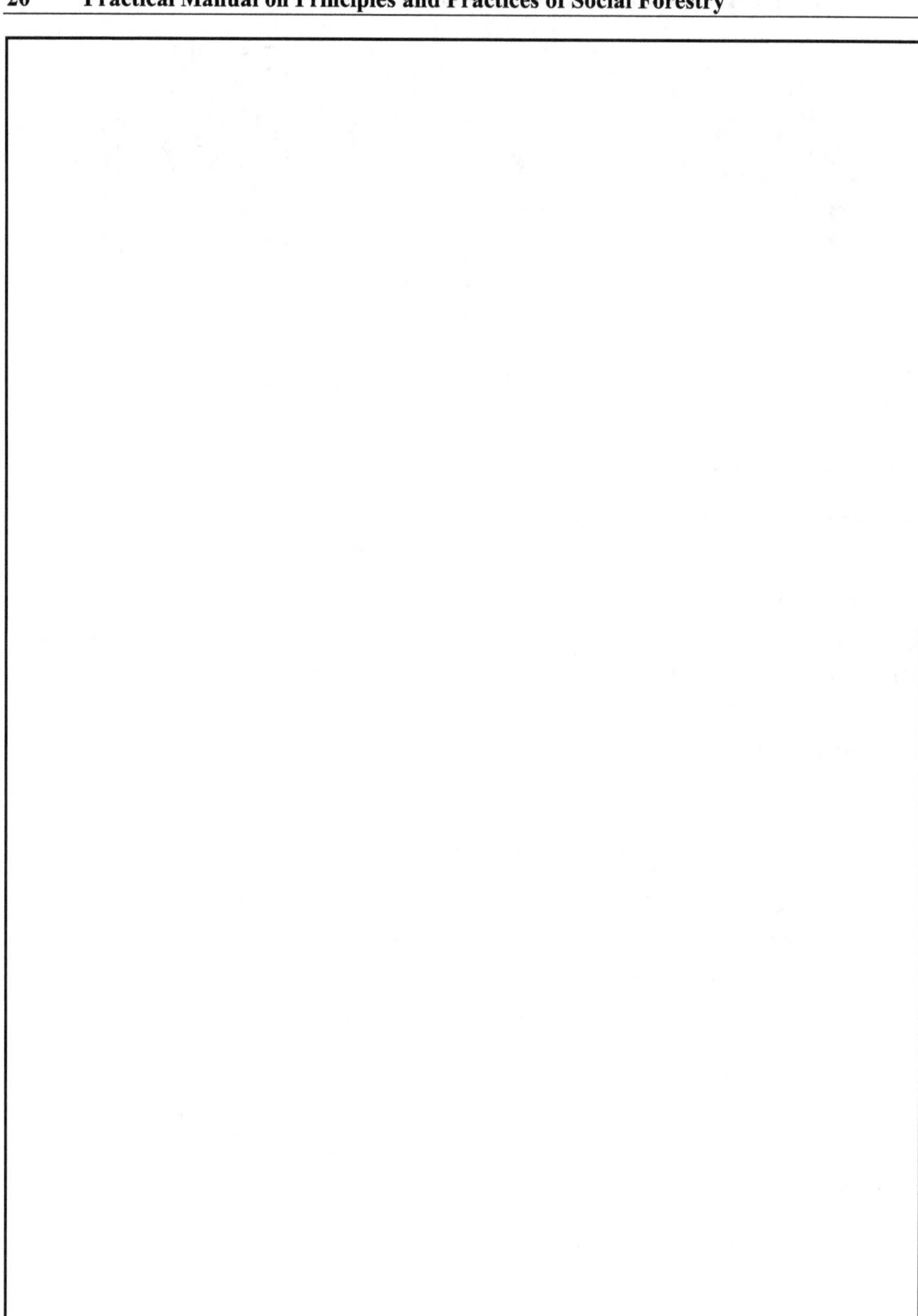

It is small to medium sized tree. It has straight cylindrical and deeply furrowed bole and a light spreading crown.

Morphological description: Leaves are feathery, paripinnate with 10-30 pairs of leaflets and oblong. Flowers are pink coloured, large and profuse. Fruit is a long pod containing about 40 seeds.

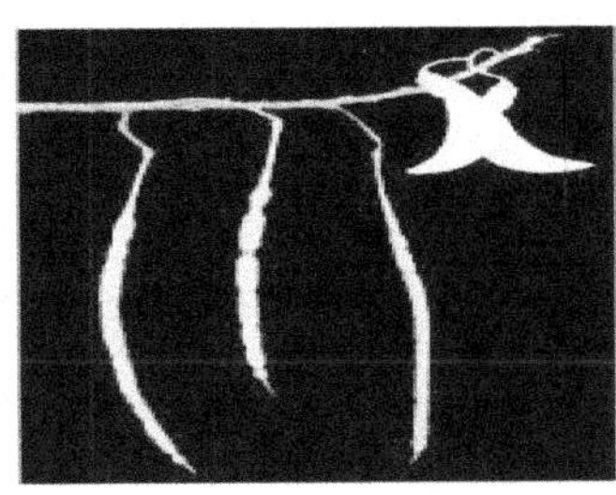

Leaf   **Twig with flowers and pods**

Economic importance: Leaves and pods are palatable fodder. It is also best used for green manuring and fuelwood.

3. Khejri *(Prosopis cineraria)*

Vernacular names: Hindi- Khajra, Telugu – Jammi.

Family: Leguminosae (Mimosoideae)

The Khejri is the prominent thorny tree of dry regions of India. It is a small to medium sized tree with much branched and dark green foliage.

Morphological description: Leaves are bipinnate with 7-10 pairs of leaflets arranged and opposite. Flowers are yellow or creamy white coloured. Fruit is a cylindrical pod globose with 10-15 seeds.

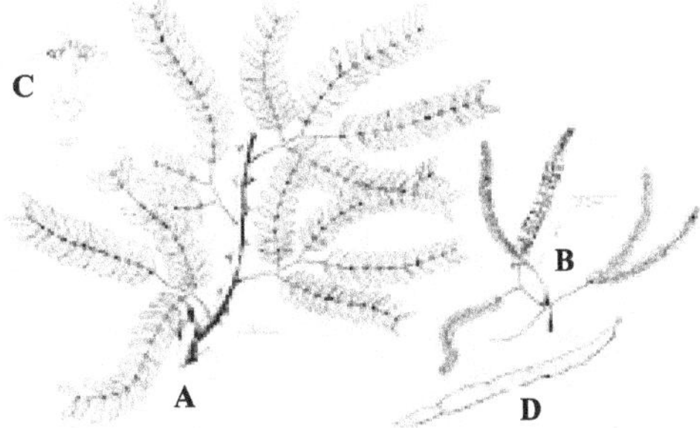

Prosopis cineraria: A: Twig B: Inflorescence C: Flower D: Fruit

Economic importance: Leaves and pods are used as fodder. The wood is hard but not durable and hence used as fuelwood.

4. Maharukh *(Ailanthus excelsa)*

Vernacular names: Hindi – Maharukh, Telugu – Peddamanu.

Family: Simarubiaceae

Ailanthus excelsa is a native of Indian Peninsula and grows almost throughout tropical and subtropical parts of country. It grows on a wide variety of soils. It is a deciduous tree.

Morphological description: Leaves are broadly sickle with 8-14 pairs of leaflets arranged nearly opposite. Flowers are yellowish, small in size and arranged in panicles. Fruit is a red, one seeded samara and winged.

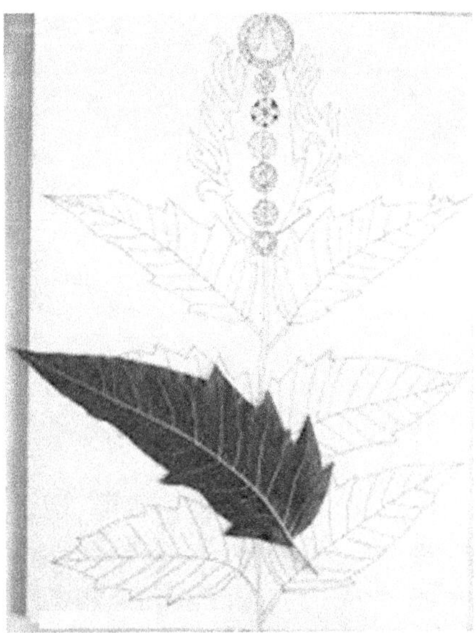

Economic importance: The leaves are rated as highly palatable and nutritious fodder for sheep and goats.

IDENTIFICATION OF TREE SPECIES SUITABLE FOR ROADSIDE PLANTATION, FIELD BUNDS, WINDBREAKS AND FOR WASTELANDS

Objectives:

1. ____

2. ____

3. ____

(A) Species for roadside plantation

1. Copper pod tree *(Peltophorum ferrugineum or Peltophorum pterocarpum)*

Vernacular names: English: Hilly tamarind, Telugu - Kondachinta, Pacha turai.

Family: Leguminosae

The tree is a native of Sri Lanka and spread into Malayan peninsula and Northern Australia. It was introduced into India at the beginning of 19th century. It is a large, evergreen avenue tree.

Morphological description: Branches are well spread. Leaves are compound, leaflets opposite, oblong, obtuse and bright green at the beginning and later turning to deep green. Flowers are bright yellow on erect rusty tomentose panicle. Fruit is copper coloured pod or rusty brown pod.

Leaves

Flowers

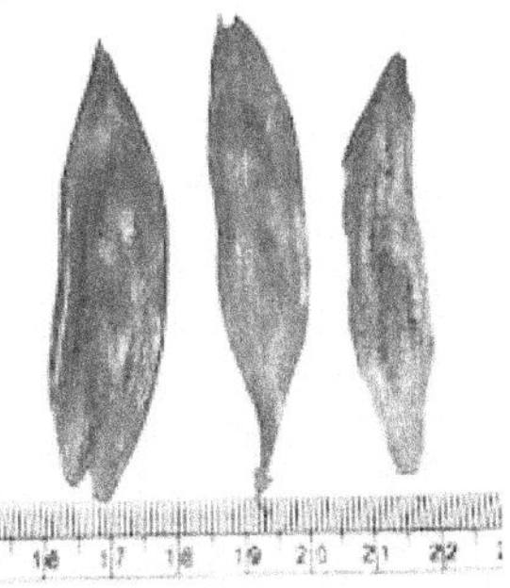

Pods

2. *Rain tree (Samanea saman)*

Vernacular names: Telugu - Nidraganneru.

Family: Leguminosae

It is a large deciduous tree with a spreading crown attaining a height of 15-25 m.

Morphological description: Leaves are compound, bipinnate and opposite, the terminal pair of leaves are larger in size. Flowers are arranged in axillary or terminal branches. Fruit is a broad pod, flat and indehiscent.

| **Tree with leafs and pods** | **Flower** | **Tree** |

3. Neem or Margosa *(Azadirachta indica)*

Vernacular names: Hindi - Nim, Telugu – Vepa.

Family: Meliaceae

Neem is said to be the special gift of nature to India. It is a large glabrous evergreen tree.

Morphological description: Leaves are bipinnate, alternate and bright green in colour, falcate, lanceolate and serrate. Flowers are white, slightly honey scented. Fruit is a fleshy drupe, turns yellow on ripening and afterwards into purple colour.

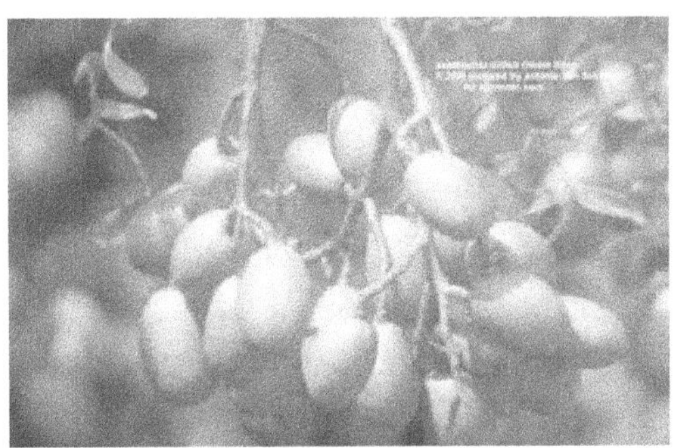

| **Leaf** | **Fruit** |

4. Tamarind *(Tamarindus indica)*

Vernacular names: Hindi - Imli, Telugu - Chinta.

Family: Leguminosae (Caesalpinaceae)

Tamarind is large, handsome and evergreen tree mostly found in South India and cultivated throughout India. But it is not originally indigenous to India and is believed to be indigenous to tropical Africa.

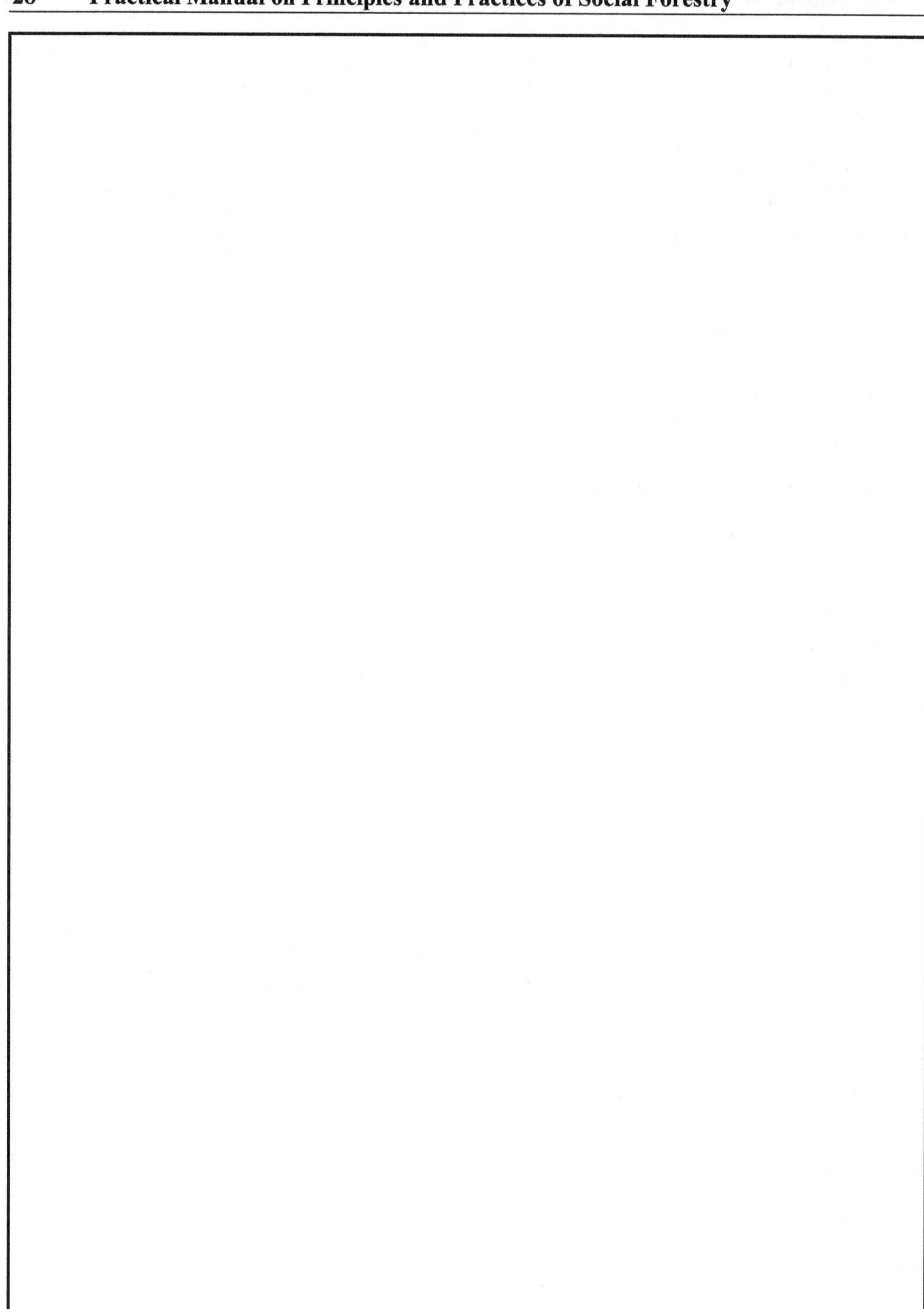

Morphological description: Leaves have 10-15 pairs of leaflets. Flower is red and yellow in colour. Fruit is thick pod filled with dark brown acid pulp and seeds are brown shining without albumen.

5. Banyan tree *(Ficus bengalensis)*

Vernacular names: Hindi – Bargad, Telugu – Marri.

Family: Moraceae

Banyan is one of the gaint trees of the world. This tree sends its prop roots from its branches continuously. The prop roots later develop into stems.

Morphological description: Leaves are ovate, mostly obtuse base, cordate or rounded, leathery, smooth and shiny in nature. Flowers are white with corymbose panicles. Fruits held together in the pouch called fig and borne from the leaf stalks and greenish in colour and turn to coral red in colour at the time of maturity.

Prop roots	**Leaf**	**Fruit**

(B.)   Species for plantation on field bunds

1. Curry leaf *(Murraya koenigii)*

Vernacular names: Telugu - Karivepaku.

Family: Rutaceae

A small pubescent tree with a short trunk, close and shady crown. It is leaf less during a short time in the hot season.

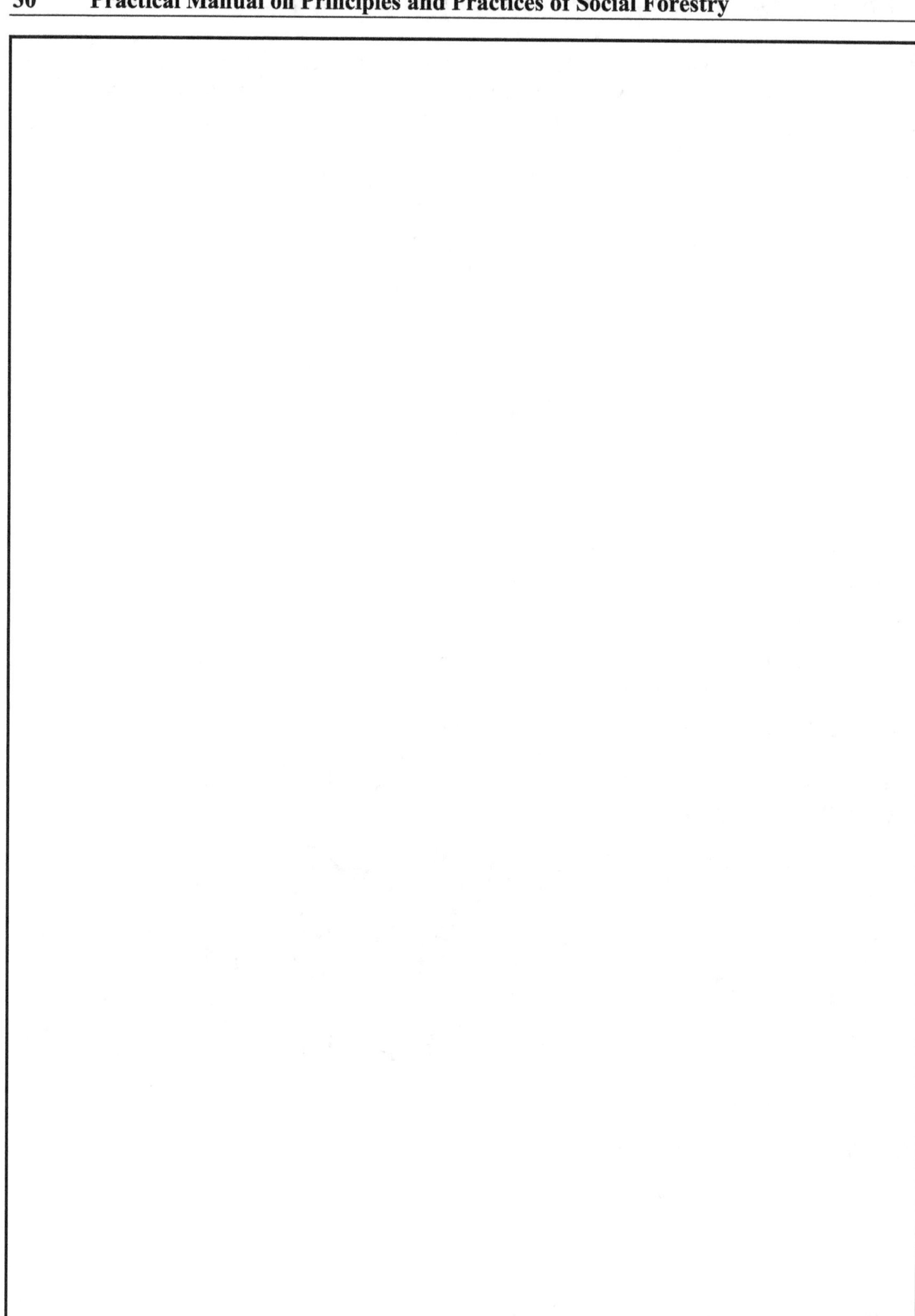

Morphological description: Leaves are with 11-25 leaflets with oblique base, ovate and lanceolate. Flower is black and rugose.

| Leaf | Flower | Fruit |

2. Drumstick *(Moringa oleifera)*

Vernacular names: Hindi – Mungna, Telugu – Munuga, English - House radish tree, vegetable tree

Family: Moringaceae

It is found wild all over India. It is a common backyard tree in every South Indian house because of its long fruits are used in every curry and soups. It is medium sized to fairly large tree with corky bark and soft spongy wood. The fruit look like sticks used for drum beating, hence called drumstick.

Morphological description: Leaves are tripinnate, alternate and light green in colour and leaflets are ovate or elliptic. Flowers are strongly honey scented with yellowish white in colour. Fruit is a long capsule, pendulous and longitudinally ribbed.

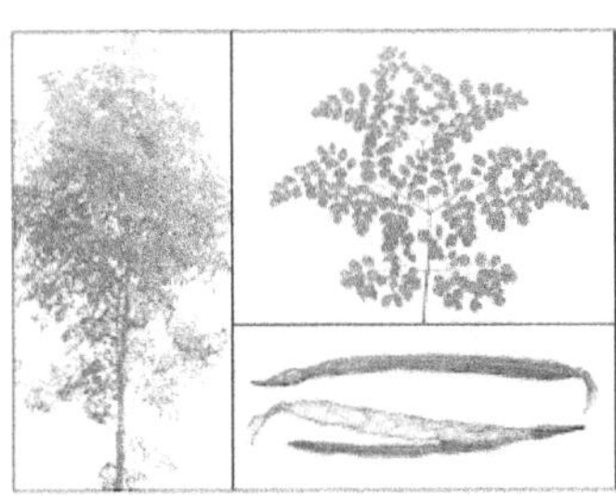

| Tree, leaf and fruit | Flowers | Fruits |

3. Gliricidia *(Gliricidia maculata)*

Vernacular names: Hindi - Vilayati shiris, Telugu – Madri.

Family: Leguminosae

The gliricidia originated from Central America and became naturalized in India during 19th century. It is a medium sized deciduous tree with soft wood and quick growth with abundant foliage.

Morphological description: Leaves are compound consisting of 8 pairs of leaflets with solitary one at the tip. Leaflets are oval and acute at the tip. Flowers are pink at early stage and turned to lilac. Fruit is a long pod with two valves looking as long flat bean.

4. Pulcherrima (*Poinciana pulcherrima*)

Family: Leguminosae

It is a shrub and shows luxuriant growth on pruning. It is an ornamental shrub.

Morphological description: Leaves are bi-compound. Flowers are with clawed petals that are yellow or golden coloured, profuse flowering during the hot rainy season and making the plant very ornamental.

Twig with leaves, flowers and fruit

(C) Species for windbreaks

1. Casuarina (*Casuarina equisetifolia*)

Vernacular names: Hindi – Jangli sari, Telugu – saru, English – Beef wood.

Family: Casuarinaceae

Casuarina is a large, fast growing tree in the graceful appearance resembling feathery conifer. The name casuarina derived from latin word "Casuaris" which means feather. The native home of casuarina is believed to be east coast of Indian subcontinent.

Morphological description: Stem is long cylindrical with many side branches. Leaves are reduced to minute scales forming whorl around the nodes. These dark green, long, slender

leaves perform the functions of leaf. Flowers are unisexual and tree flowers twice in a year during April-May and again in Sept-October. Fruit is a woody globose cone.

Twig with leaves and fruit	**Globose cone fruit**

2. Eucalyptus sps. *(Eucalyptus camaldulensis, Eucalyptus teriticornis, Eucalyptus hybrid etc.)*

Vernacular names: English: Mysore gum, Telugu – Euclis oil or Jama oil, Nilgiri.

Family: Myrtaceae

Eucalyptus is introduced from Australia during 18th century as an ornamental tree. It is a tall, elegant, straight and clean boled tree.

Morphological description: Leaves are thin and green, leaves of sapling generally opposite, sessile, cordate and horizontal and those of adult trees are alternate, petiolate and vertical. Flowers are in umbels or heads and pedunculate. Fruit is usually woody with full resin sacs.

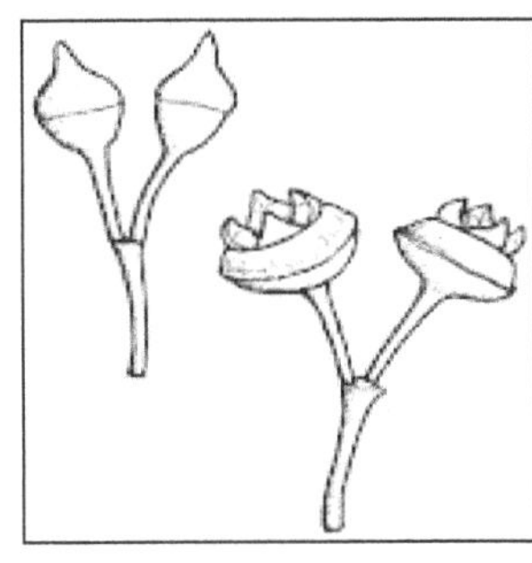

Leaf	**Fruit**

(D)   Species for wasteland afforestation

1. Indian coral tree (*Erythrina indica* or *Erythrina variegate*)

Vernacular names: Hindi – Dadap, Telugu – Badish.

Family: Leguminosae

Indian coral tree is one of the loveliest trees found in the deciduous forests. It is small to medium sized tree and attains a height of more than 15 meters under favourable conditions.

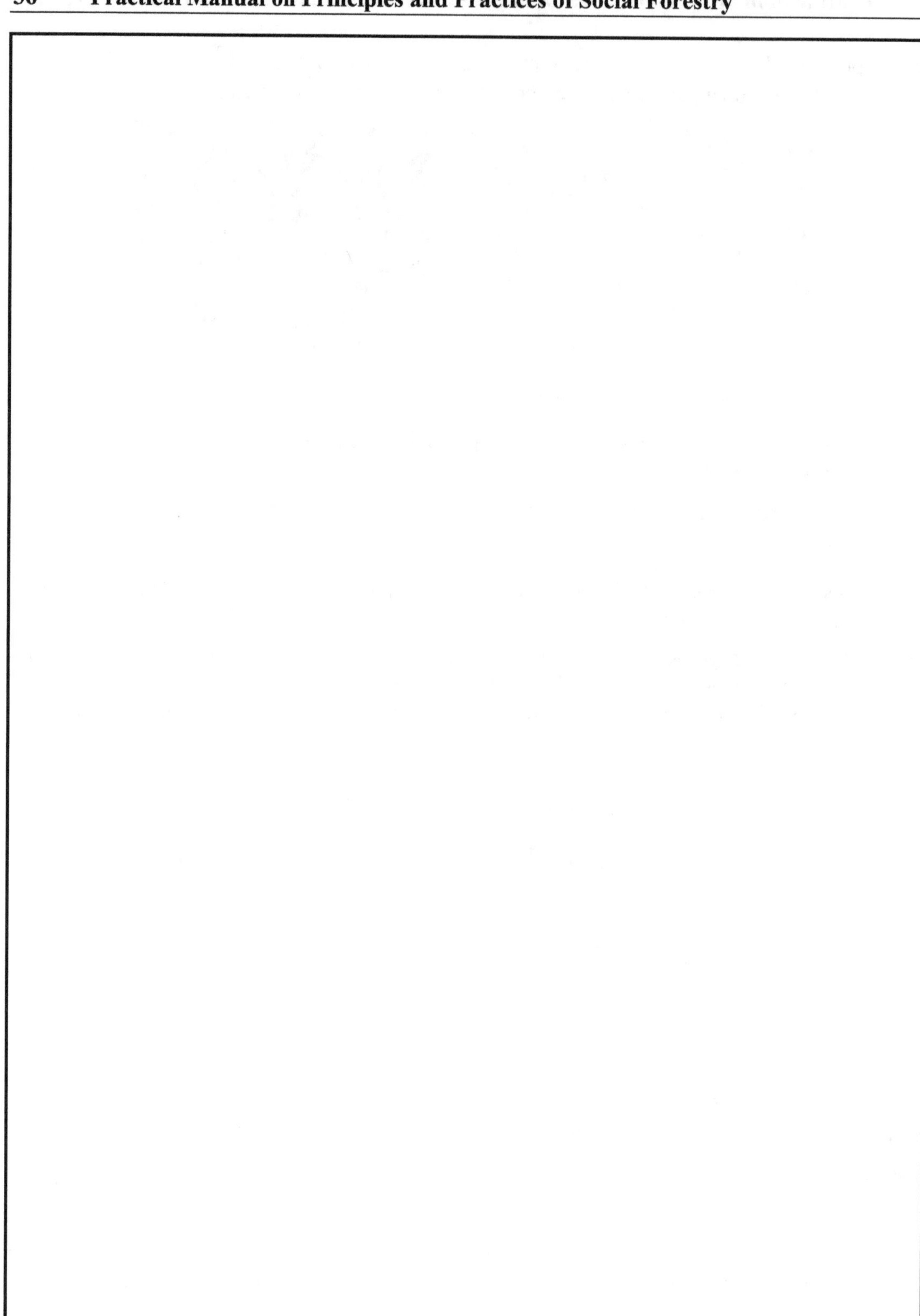

Morphological description: Leaves are large and thick. Each leaf has 3 leaflets with long stalk and glabrous. Flowers are bright scarlet in dense racemes. Fruit is stout and greenish pod but later turned to deep brownish colour.

Leaf	**Flower**	**Fruit**

2. Pongam oil tree (*Pongamia pinnata*)

Vernacular names: Hindi – Karanj, Telugu – Ganuga, Pungu, English – Indian Beach.

Family: Leguminosae (Papilionaceae)

It is a native of Indian sub continent. It is a medium sized glabrous, nearly evergreen tree with a spreading crown and short bole.

Morphological description: Leaves are imparipinnate, glabrous, bright green and the compound leaf consists of 5-9 leaflets that are present in pairs. Flowers are raceme and display various colours. Purplish or lilac colour flowers are common but occasionally white flowers are seen. Fruit is indehiscent pod and turgid and contain an oblong single seed.

Leaf	**Fruit**	**Seed**

3. Hardwickia (*Hardwickia binata*)

Vernacular names: Hindi – Anjan, Telugu - Yepi, Narayepi.

Family: Leguminosae

Thick is a tree of dry hot climate. It grows on a wide variety of rocks and soils. It is a large gregarious tree and deciduous for a short period towards the end of the cold season.

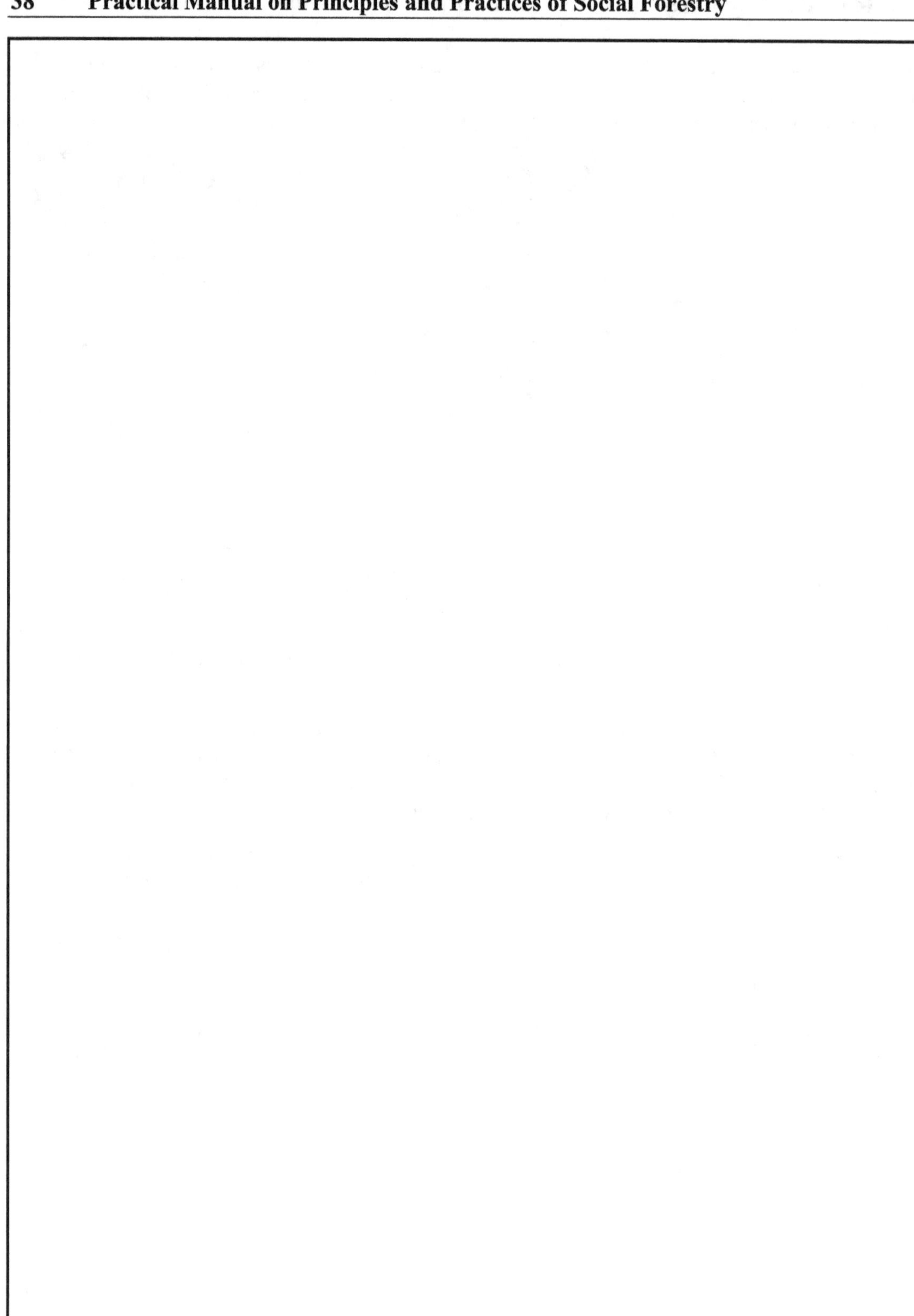

Morphological description: Leaves are with 2 leaflets, obtuse, obliquely ovate to trapezoid, flowers are pale yellowish green, appear in axillary and terminal lax panicle racemes. Fruit is flat pod, oblong, lanceolate, coriaceous with parallel longitudinal veins.

Leaf

leaves and fruits

4. Flame of the forest *(Butea monosperma)*

Vernacular names: Hindi – Dhak, Palas, Telugu – Moduga.

Family: Leguminosae

Palas is indigenous medium sized tree of India. The trunk and branches are somewhat crooked and covered with grayish coloured bark.

Morphological description: Leaves are trifoliate. Leaflets are moraceous, hard, ovate obtuse. Flowers are large and crowded, rich orange colour. Fruit is a flat pod and dry pods hang down even after drying.

Leaf

Flower

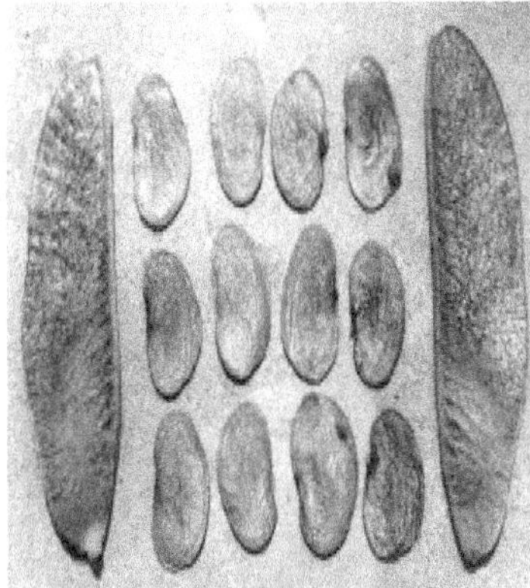

Pod and seeds

5. Bauhinia *(Bauhinia variegate, Bauhinia purpurea)*

Vernacular names: Hindi – Kachnar, Telugu – Kanchan.

Family: Leguminosae

It is a medium sized tree growing all over India. The tree is indigenous of India and fairly common in deciduous forests.

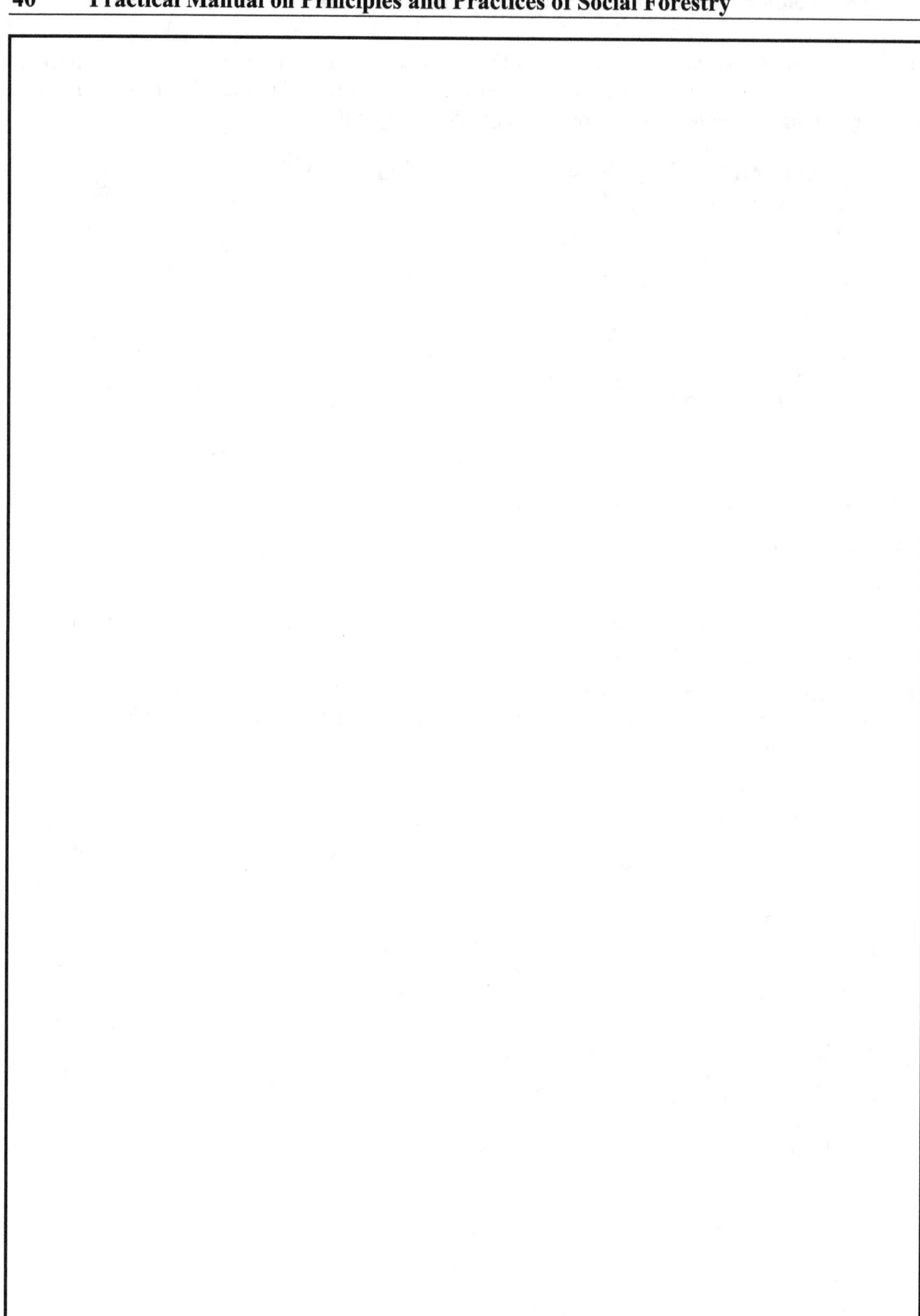

Morphological description: Leaves are more or less oval shaped and split at the tip into two lobes, coriaceous, base of the leaf is rounded and the tips are slightly pointed. Flowers are purple or white and sometimes both colours are seen on the same tree, racemes short in terminal. Fruit a long pod, dehiscent, hard and flat.

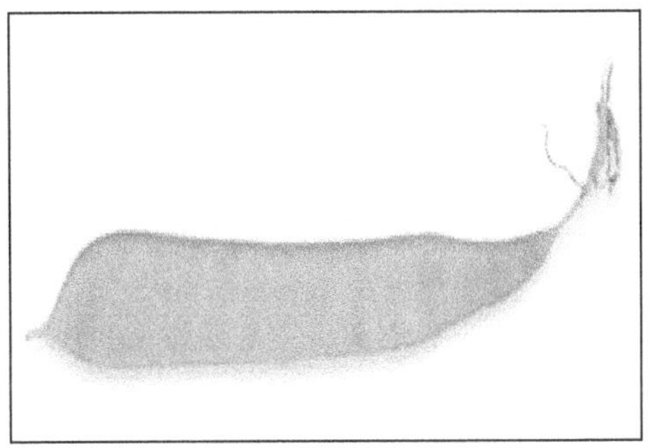

| Leaf | Flower | Fruit |

6. India laburnum (*Cassia fistula*)

Vernacular names: Hindi – Amalta, Telugu – Rela.

Family: Leguminosae

Its origin is from Indian sub-continent. It is found abundant in dry and moist deciduous forests of India. It is a medium sized and slow growing tree.

Morphological description: Leaves are compound and rich green in colour. Each leaf possesses 4-8 leaflet pairs, glabrous, acute or shortly acuminate. Flowers are long droopy raceme, large, bright yellow and attractive. Fruit is cylindrical pod, long and black in colour.

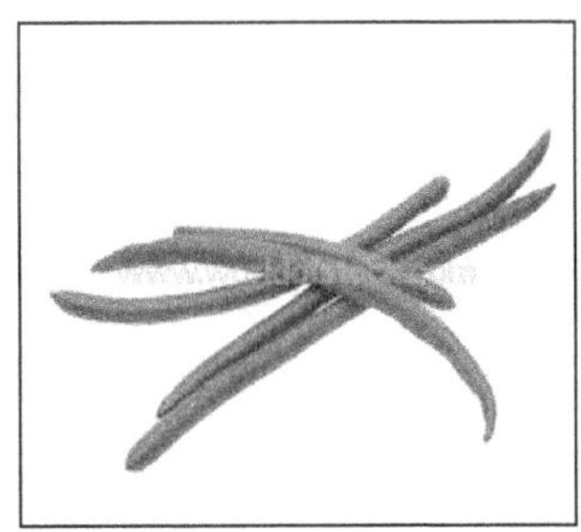

| Leaf | Drooping flower | Fruit |

7. Bamboo (*Dendrocalamus strictus*)

Vernacular names: Hindi – Bans, Telugu – Veduru.

Family: Gramineae (Bambuseae)

Bamboo is a gaint tropical grass that is abundantly grown in peninsular India. It is middle sized deciduous tall grass. It is commonly cultivated throughout India in the plains and foothills.

Morphological description: It is gaint tall grass consists of narrow green leaves. Each clump contains 2 to 50 shoots called culms. The culms grow high rapidly but they are close together at the bottom. The culms are attractive green at the young stage and later turned to brown colour after drying. The culms are almost solid in dry area and in a moist climate they are hollow with thick walls. The flowering of this species occurs both sporadically almost every alternate year and also gregariously at long intervals of 20-65 years.

Leaf

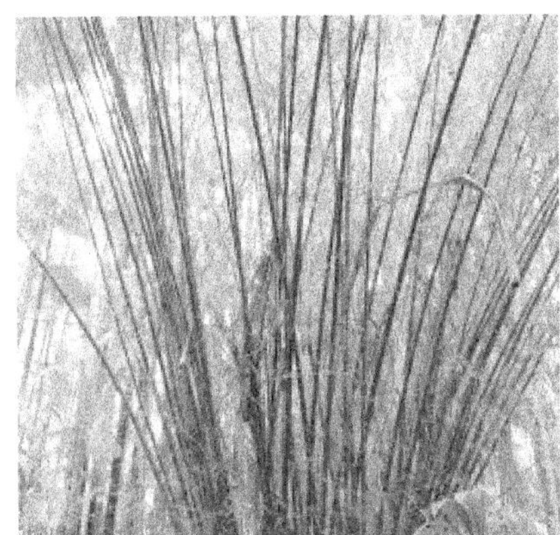

Culms

IDENTIFICATION OF MINOR FOREST TREE SPECIES, TREES FOR BEAUTIFICATION PURPOSE AND NITROGEN FIXING TREE SPECIES AND OTHER SPECIES SUITABLE FOR AGROFORESTRY

Objectives:

1. ____

2. ____

3. ____

(A)   Minor forest trees

1. Ber (*Zizyphus maruritiana*) (*Zizyphus jujube*)

Vermacular names: Hindi - Pitniber, Telugu-Regu, English-India jujube.

Family: Rhamnaceae

The ber is said to be indigenous to India. It is a large shrub that is found in dry deciduous forests.

Morphological description: Leaves are thick variable from ovate-oblong to nearly orbicular, obtuse or acute, entire or serrulate. Flowers are thick greenish yellow in short axillary nearly sessile cymes. Fruit is crisp drupe and light yellowish scarlet in colour.

2. Soapnut (*Sapindus laurifolius*) (*Sapindus trifoliatus*)

Vernacular names: Hindi-Rita, Telugu- Kunkudu.

Family: Sapindaceae

The tree seems to be a native of India. It is well known tree in South India because of its extensive use for hair washing. It is medium sized tree.

Morphological description: Leaves are compound consisting of 2-3 pairs of leaflets and largest pair found at the terminal point, elliptic and generally obtuse. Leaves are rusty and pubescent. Fruit is coriaceous or fleshy. Pericarp contains **saponine** which makes lather with water.

3. Jamun or Jambul (*Syzygium cumini*)

Vernacular names: Hindi – Jamun, Telugu- Neradu, English-Black plum, Indian Black berry

Family: Myrtaceae

Jamun is said to be indigenous to India and is found wild in the forests of tropical and sub tropical parts of the country. The tree is evergreen large sized tree.

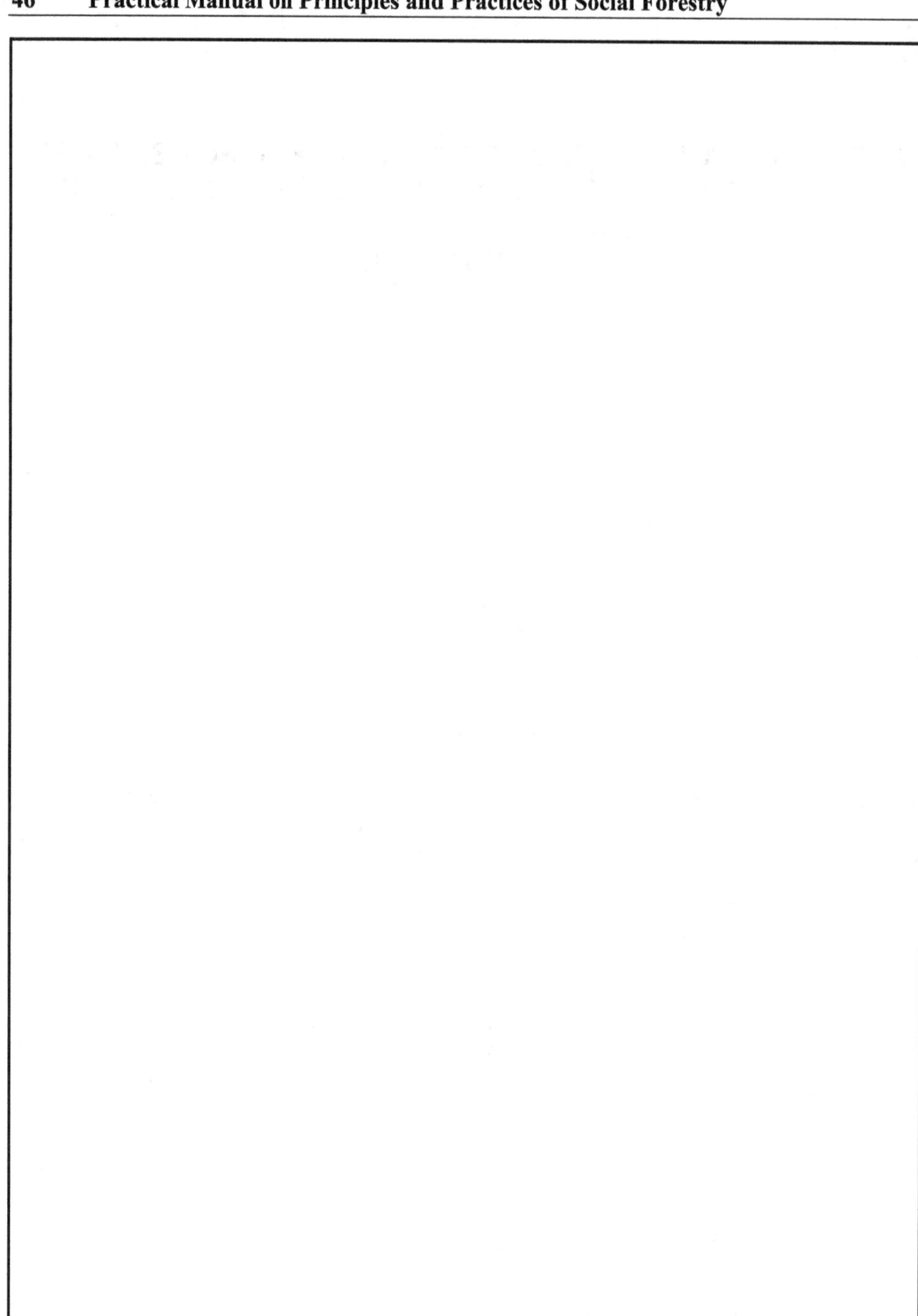

Morphological description: Leaves are coriaceous, elliptic and lanceolate. Leaves are whitish, scented and sessile. Fruit is green at early stage but turned to purple later and shiny, attractive, sweet and sub-acidic in taste.

4. Khair (*Acacia catechu*)

Family: Leguminosae

It is a medium sized tree with dark trunk. It is used for medicinal, fuel and fodder for goat and sheep, mainly used for gum katha and dye.

Morphological description: Leaves are pinnate with 10 -12 pairs. Leaflets 30-50 pairs, linear, glabrous. Leaves are pale yellow with cylindrical spikes. Fruit is thin pod, brown, shining and dehiscent.

5. Tendu (*Diospyros melonoxylon)*

Vernacular name: Hindi-Tendu, Telugu-(Beedi leaf tree).

Family: Ebenaceae

Tendu is a moderate, deciduous tree and occurs widely in mixed deciduous forests of Madhya Pradesh, Andhra Pradesh, Maharastra, Bihar and Orissa.

Morphological description: Leaves are long, elliptic, lanceolate and alternate, hairy when young and glabrous after maturity.

Flower: Male flowers are sub sessile in short dropping cymes, female flowers are solitary and larger than male flowers. Fruit is a berry, globose, glabrous, olive green but turns to yellow when ripe.

6. Chironji (*Buchanania latifolia* or Syn. *Buchanania lanzan*)

Vernacular names: Hindi - Chiroli, Telugu - Sara.

Family: Anacardiaceae

Buchanania latifolia is a common tree in deciduous forests of India. It is a moderate sized tree with a short trunk.

Morphological description: Leaves are coriaceous, hard, oblong and obtuse. Flowers are sessile and greenish white in colour. Fruit is a drupe, black in colour when ripe and encloses a hard bony stone.

7. Sandal wood (*Santalum album*)

Vernacular names: Hindi - Chandan, Telugu- Chandanam or Gandham.

Family: Sanatalaceae

Sandalwood is a small, evergreen and glabrous tree with slender dropping branchlets. Sapwood is white and scentless and heartwood is yellowish brown and strongly scented.

Morphological description: Leaves are opposite, ovate, lanceolate. Flowers are brownish purple in colour in axillary or terminal. Fruit is drupe, globose and black with hard endocarp.

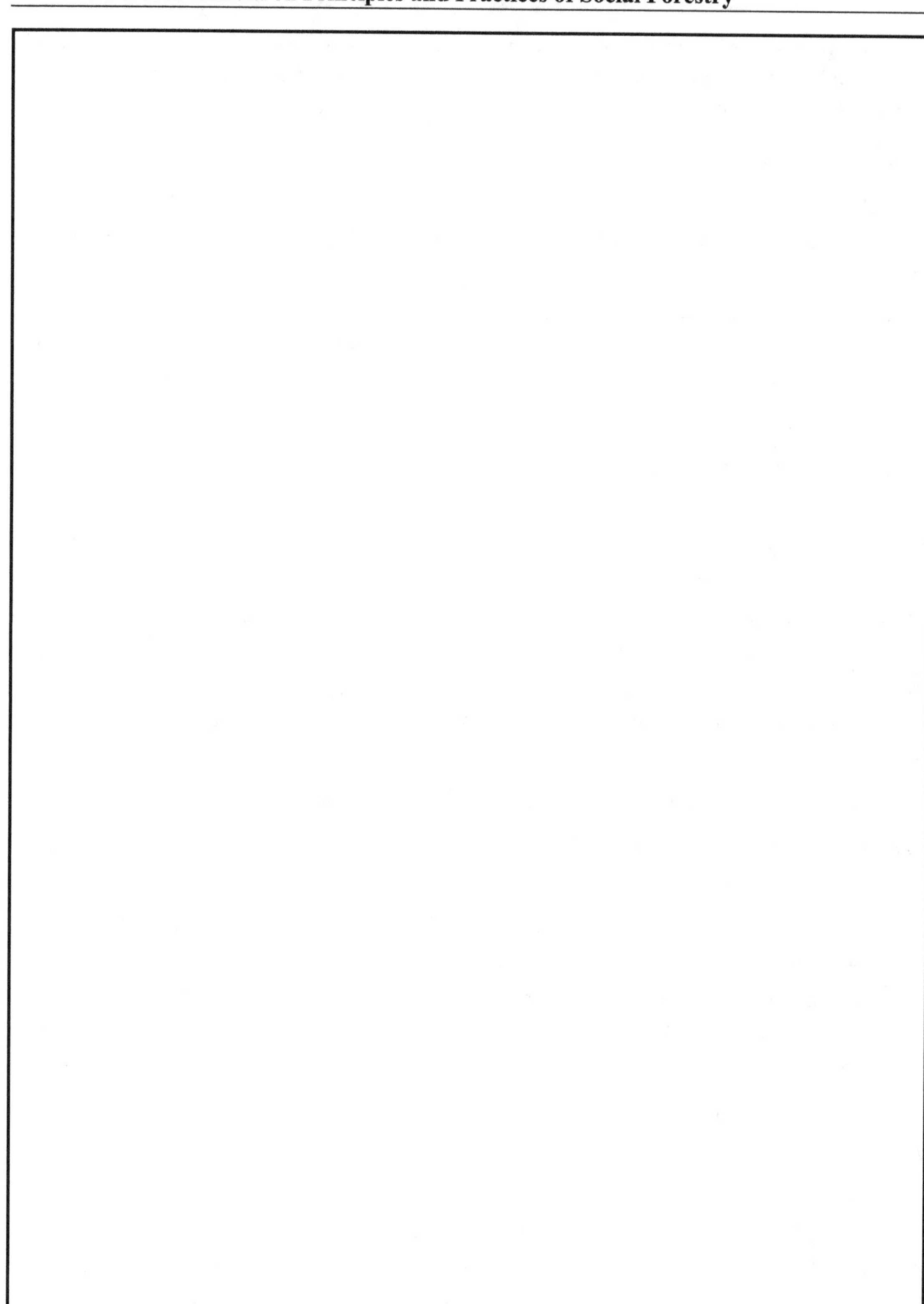

(B)   Trees for beautification purpose

1. Gulmohar (*Delonix regia*)

Vernacular names: Telugu – Shima sunkesula or Erra turai.

Family: Leguminosae (Caesalpinaceae)

It is a spreading like tree and native of Madagascar. It is a very beautiful ornamental tree with a brilliant mass of scarlet flower and look like the red flame on the canopy.

2. African Tulip tree (*Spathodea companulata*)

Vernacular names: Telugu – Patadiya.

Family: Bignoniaceae

It is a tall growing and ornamental tree with orange crimson flowers and is one of the finest trees for scenic planting. The straight trunk and abundant long lasting flowering makes the tree most important among the beautiful trees.

3. Cork tree (*Millingtonia hortensis*)

Vernacular names: Telugu - Kada malli.

Family: Bignoniaceae

It is a tall, straight and very beautiful ornamental tree. It is also known as Akash nim and Nim chaveli. The tree bears numerous white flowers and scented.

4. Jacaranda (*Jacaranda mimosifolia*)

Family: Bignoniaceae

It is also beautiful ornamental tree and is grown as avenue tree. The tree is small sized with beautiful flowers in blue violet colour.

5. Indian Tulip tree (*Thespesia populanea*)

Vernacular names: Hindi – Bhendi, Telugu – Ganga ravi.

The tree is indigenous of Indian subcontinent and grows wildly. It is a medium sized evergreen tree. Flowers are showy and bright yellow with a purple eye in the side at base. The old flowers turn to brick red or deep maroon before they fall off.

6. Mast tree (*Polyalthia longifolia*)

Vernacular names: Hindi – Ashoka (misnomer), Telugu – Narmamidi, English – Indian fir.

Family: Anonaceae

The mast tree is indigenous of peninsular India and distributed throughout the country. Mast tree is a tall, elegant, middle sized and almost evergreen tree. The leaves are long, lance shaped, shiny and dark green. Greenish flowers are grouped into bundles. It is a very beautiful ornamental tree.

Other ornamental tree

1. *Tecoma argentia*   (Family – Bignoniaceae)
2. *Tecoma stans*   (Family – Bignoniaceae)
3. *Legostromia floregenea*   (Family – Leguminosae)
4. *Cassia marginata*   (Family – Leguminosae)
5. *Saraca indica*   (Family – Leguminosae)

(C)   Nitrogen fixing tree species suitable for agroforestry

1. *Acacia albida (Faidherbia albida)*
2. *Leucaena leucocephala*
3. *Gliricidia maculata*
4. *Erythrina indica*
5. *Sesbania grandiflora*
6. *Pongamia pinnata*
7. *Dalbergia sissoo*
8. *Albizzia lebbek*
9. *Casuarina equisetifolia*

(D)   Other species (Non - Nitrogen fixing trees (NFT) suitable for agroforestry

1. *Morus alba*
2. *Emblica officinalis*
3. *Melia azaderach*
4. *Bombax ceiba*
5. *Sterculia foetida*
6. *Azadirachta indica*
7. *Anacardium occidentale*

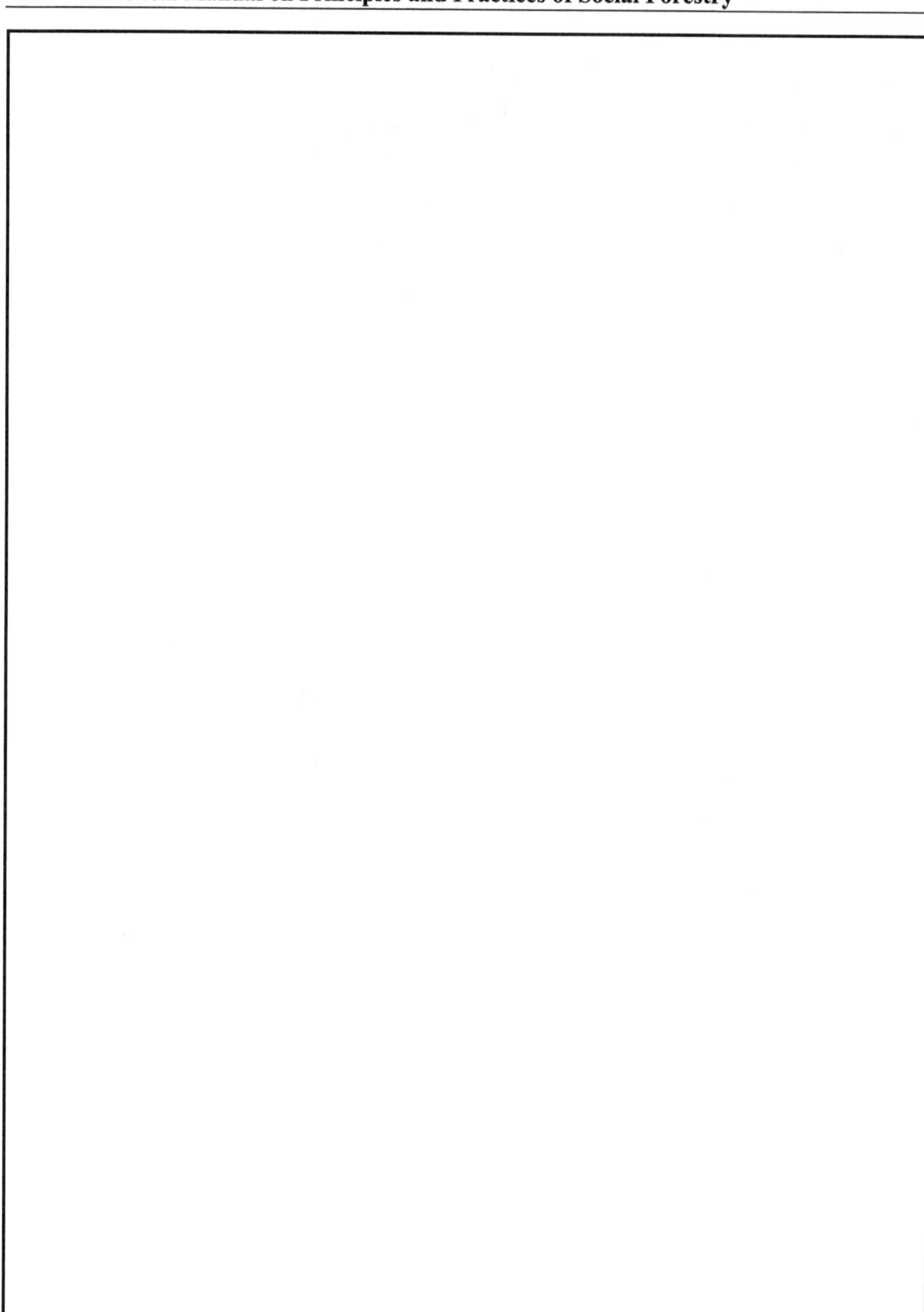

IDENTIFICATION OF SEEDS OF IMPORTANT TREE SPECIES

1. *Acacia nilotica*: Compressed and oval to elliptic, black to greenish brown, 4-12 mm long. All species have the faces marked by an oval or elliptic line more or less concentric with the outline of the seed. Endosperm lacking.

2. *Prosopis juliflora:* Compressed and oval, ovate or elliptic, 2.5-7 mm long brown, glossy, with a central ring on each face. Cross sections show the yellowish embryo sandwitched between 2 hard glassy layers of endosperm.

3. *Cassia siamea:* Diverse externally and internally, medium to large, compressed oval, flattish circular, flattish – spatulate or rhomboid. All species have faced marked by an elliptic to circular zone or ring.

4. *Sesbania grandiflora*: Oblong-ellipsoid and slightly compressed 4 × 2.5 × 2 mm, blackish, mottling on gray or brown.

5. *Hardwickia binata:* The seed is ex-albuminous, flat, 2 cm × 1 cm sub-reniform, pointed at one end and rounded at the other, with a fairly hard testa.

6. *Cassia fistula*: The seeds are 1 cm × 0.75 cm, ovate, compressed light brown, hard, smooth, shiny with a moderately hard testa and albumen.

7. *Bauhinia purpurea* (**Bahunia**): The seeds are 1.5 cm × 1.5 cm, nearly circular flat, brown with a somewhat coriaceous testa.

8. *Albizzia lebbek*: The seeds are 0.75 – 1.0 cm × 0.5 – 0.75 cm, obovate or oblong, compressed, light brown, smooth with a hard testa.

9. *Syzygium cumini*: Seeds are oblong ovoid 1-2 cm long, 2-5 angular and irregularly shaped, seeds are compressed together into a mass resembling a single seed, the whole enclosed in a sub-coriaceous covering.

10. *Diospyros melonoxylon* (**Beedi seed**) The seeds are oblong, compressed 1.25-2.0 cm long, brown, shining with a wrinkled testa and ruminate albumen.

11. *Tectona grandis* (**Teak**): Fruit is a hard, bony, irregularly globose nut somewhat pointed at the apex, enclosed in a thick, light brown covering, usually 1-1.5 cm in diameter, but varying much in size, containing 1-3 rarely 4 seeds.

12. *Bombax ceiba*: The seeds are 0.4 to 0.6 cm long, irregularly obvoid, dark brown, smooth with a brittle testa.

13. *Ailanthus excelsa*: The fruit is a red one-seeded samara, 5-8 cm long by 1-1.5 cm wide prominently veined, acute at both ends and twisted at the base.

14. *Sapindus emarginatus* (**Soapnut**): The seeds are smooth, black, nearly globose, about 1 cm long, very hard.

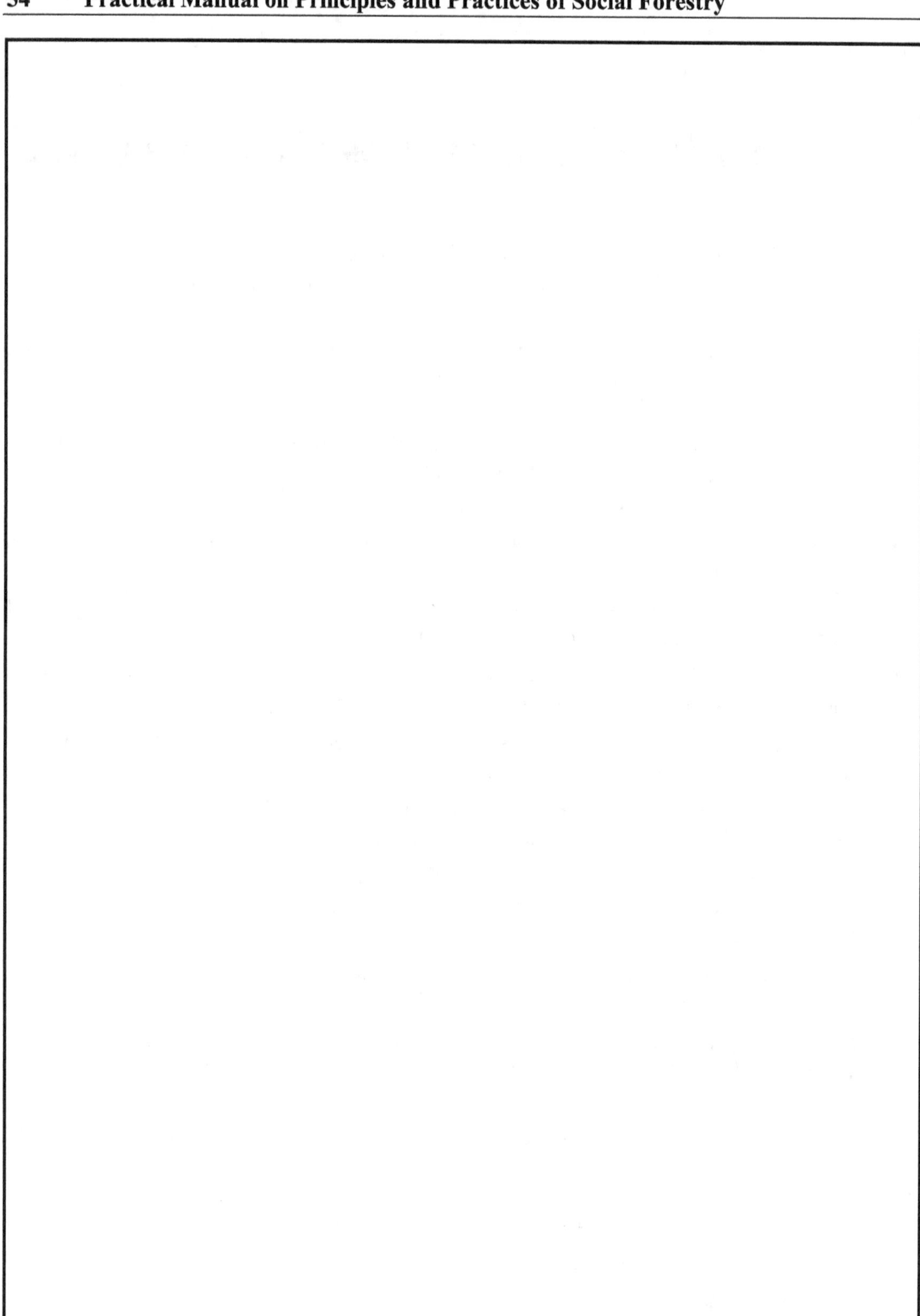

15. ***Butea monosperma*:** The seed is flat, reniform 3-4 cm × 2 - 2.5 cm with a thin papery reddish brown testa.

16. ***Dalbergia sissoo* (Sissoo):** The seeds are 0.5-0.75 cm × 0.2 - 0.25 cm reniform, flat light brown, with a delicate papery testa.

17. ***Santalum album* (Sandalwood):** The fruit is a purplish black globose succulent drupe 1-1.5 cm in diameter, with a brown endocarp which is moderately hard but brittle and easily broken. The fruit stones are globose 0.50 – 0.75 cm in diameter. The seed has copious white albumen.

18. ***Emblica officinalis* (Aonla):** The fruit contains 4-6 dark brown smooth seeds that can be extracted by placing the ripe fruits in the sun until the hard putamen dehisces and the seeds escape.

19. ***Dendrocalamus strictus* (Bamboo):** Seeds grain like, characteristic of graminae family.

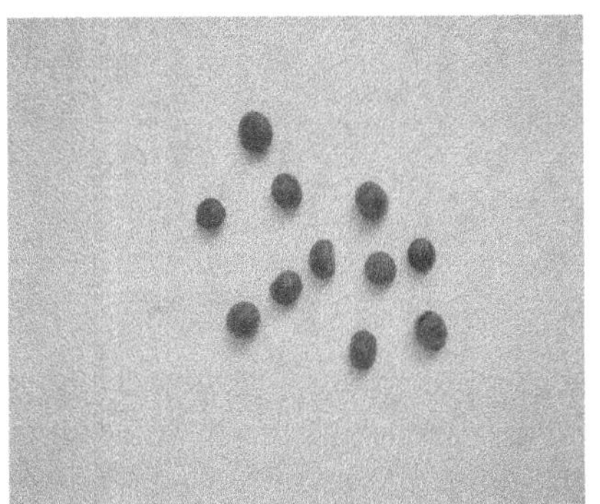

Acacia nilotica

Albizzia lebbek

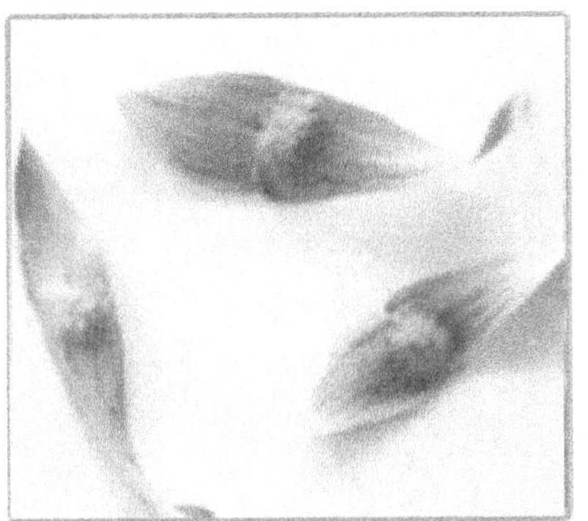

Winged fruit with a solitary seed at the center

Ailanthus excelsa

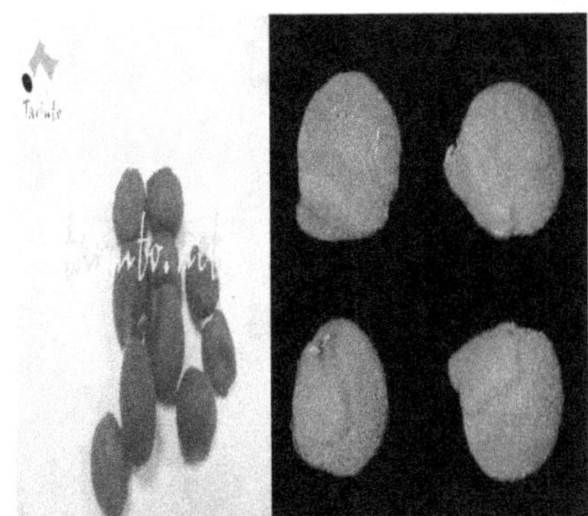

Bombax ceiba

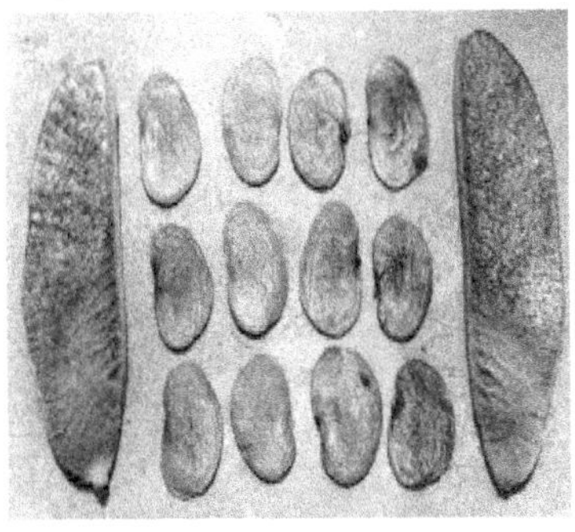

Butea monosperma

Cassia fistula

Bauhinia purpurea

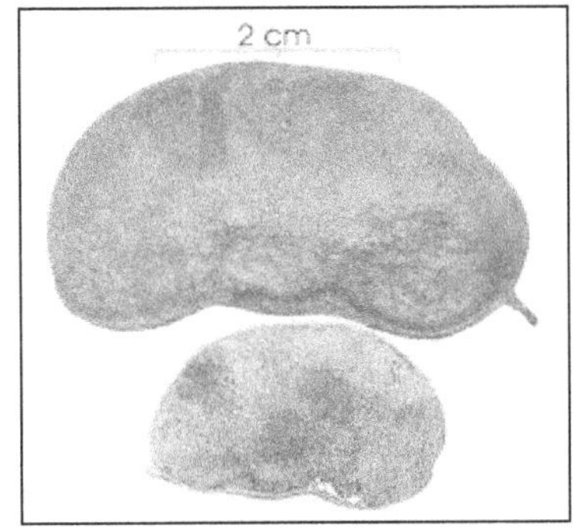

Dalbergia sissoo

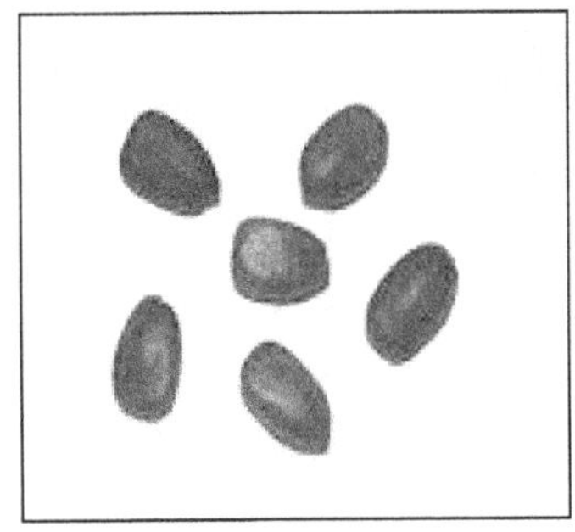

Cassia siamea

Dendrocalamus strictus

Emblica officinalis

Hardwickia binata

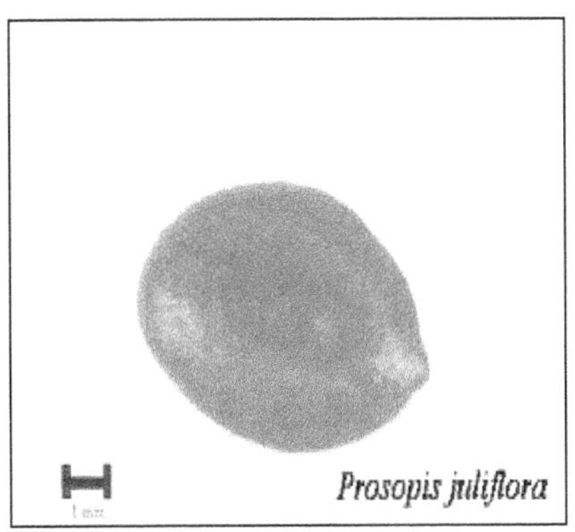

Prosopis julifera

Santalum album

Sapindus emarginatus

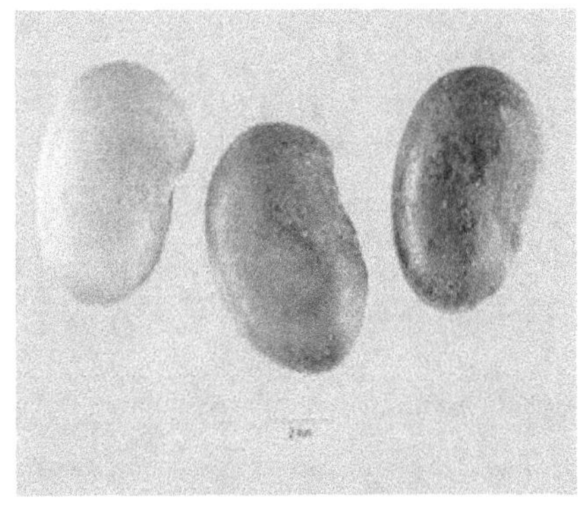

Sesbania grandiflora

Syzygium cumini

Teak

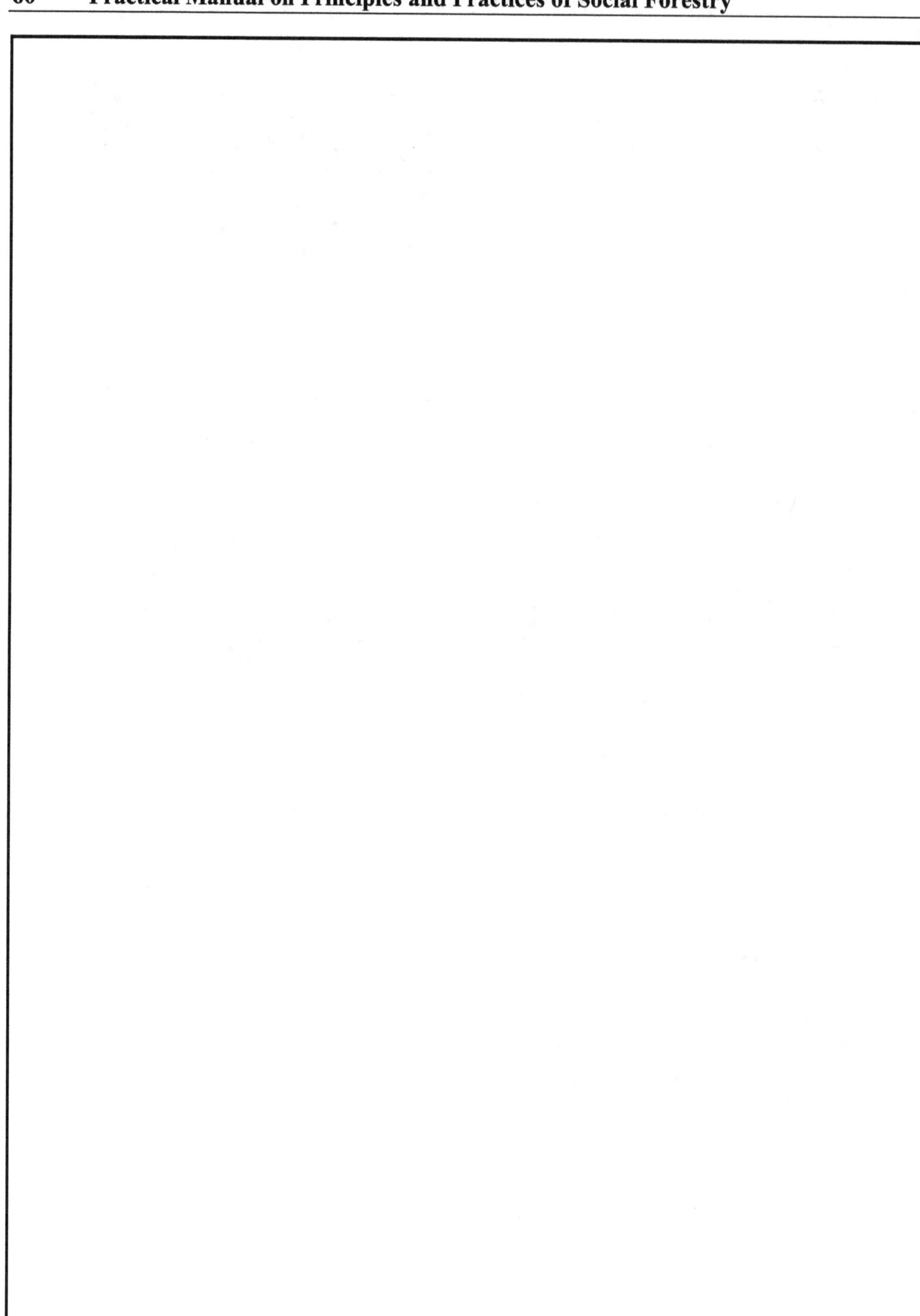

COLLECTION, EXTRACTION AND STORAGE OF TREE SEEDS

Objectives:

1. ____

2. ____

3. ____

Period of flowering and fruiting

A sufficient knowledge of the season of flowering and fruiting, interval between flowering and fruiting period, when fruit and the seed ripens, the seeding cycles and seed behaviour of important species is essential for successful implementation of planting programme.

The period of flowering and fruiting and seed ripening varies considerably with locality. There may be a difference of few months in the fruiting of the same species and therefore knowledge of the local behaviour of the species is important.

The approximate period of flowering and fruiting is indicated in the table below of some important species that may provide guidance for the seed collection work.

Species	Period of flowering	Period of ripening of fruit / seed	Remarks
Acacia nilotica	June-Sept.	April-June	
Ailanthus excelsa	February – March	May – June	
Albizzia lebbek	April – June	January – February	
A. procera	June – September	February – May	
Anacardium occidentale	December – April	March – June	
Azadirachta indica	March – May	June – August	
Bombax ceiba	January – February	April – May	
Cassia fistula	April – June	March – April	
C. siamea	March – June	March – April	
Casuarina equisetifolia	Feb - April	June – July	(Flowering & Fruiting twice in a year)
Cedrus deodara	September – Oct.	Oct – Nov.	
Dalbergia latifolia	--	March – April	
Dalbergia sissoo	March – April	Dec – Jan	
Dipterocarpus alatus	Feb. – March	May – June	
Emblica officinalis	March – May	Nov – March	

Contd....

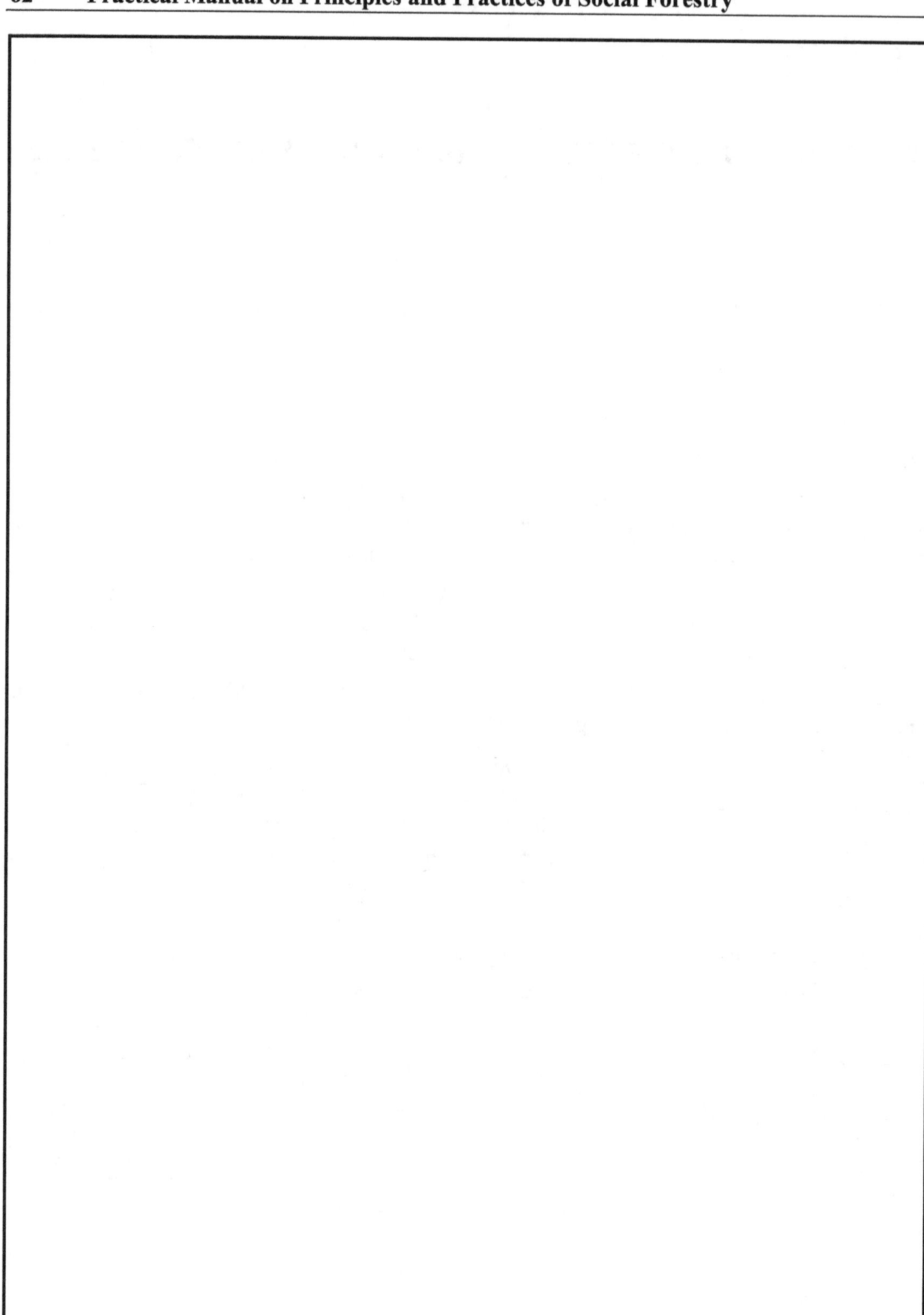

62   **Practical Manual on Principles and Practices of Social Forestry**

Eucalyptus citriodora	May – June	Jan – Feb	
E. globules	Feb – March	March – April	
Gmelina arborea	Feb – April	April – July	
Grevellia robusta	March – May	May – June	
Delonix regia	March – April	Sept – Oct	
Lagerstrominea flosreginae	April – June	Nov – Jan	
Madhuca latifolia	March – April	June – August	
Mangifera indica	March – April	May – July	
Michelia champaca	July – August	Aug. Sept.	
Melia azaderach	April – June	Jan. – Feb	
Morus alba	March – April	May – June	
Pinus roxburghii	Feb. – April	March–May (next year)	
Polyalthia longifolia	Feb. – April	July – Aug.	
Sterculia alata	Feb – March	Dec. – Feb	
Shorea robusta	March – April	July – July	
Syzygium cumini	March – May	June – August	
Tamarindus indica	April – June	Feb. – April	
Terminalia tomentosa	June – July	Feb. – March	
Tectona grandis	June – Sept.	Nov. – Jan	

Collection, Extraction and Storage of Seeds

Seed Collection

(a) Objectives of seed collection

1. To meet continuous short and long term supply of reproductive material for plantation programme *i.e.,* utilization.

2. To ensure supply of reproductive material for scientific trails and introduction of species in exotic areas *i.e.,* evaluation.

3. To supply the seed for establishment of gene-banks, expansion of arboretums, botanical gardens and seed herbaria *i.e.,* conservation.

(b) Factors effecting availability of good seed

1. Location of the seed stand.
2. Age characteristic of the tree in the stand.
3. Time of collection.
4. Seed bearing capacity of crop.
5. Methods of collection.
6. Methods of treatment of fruits and seed after collection.
7. Periodicity of fruiting.
8. Quick and safe transport.

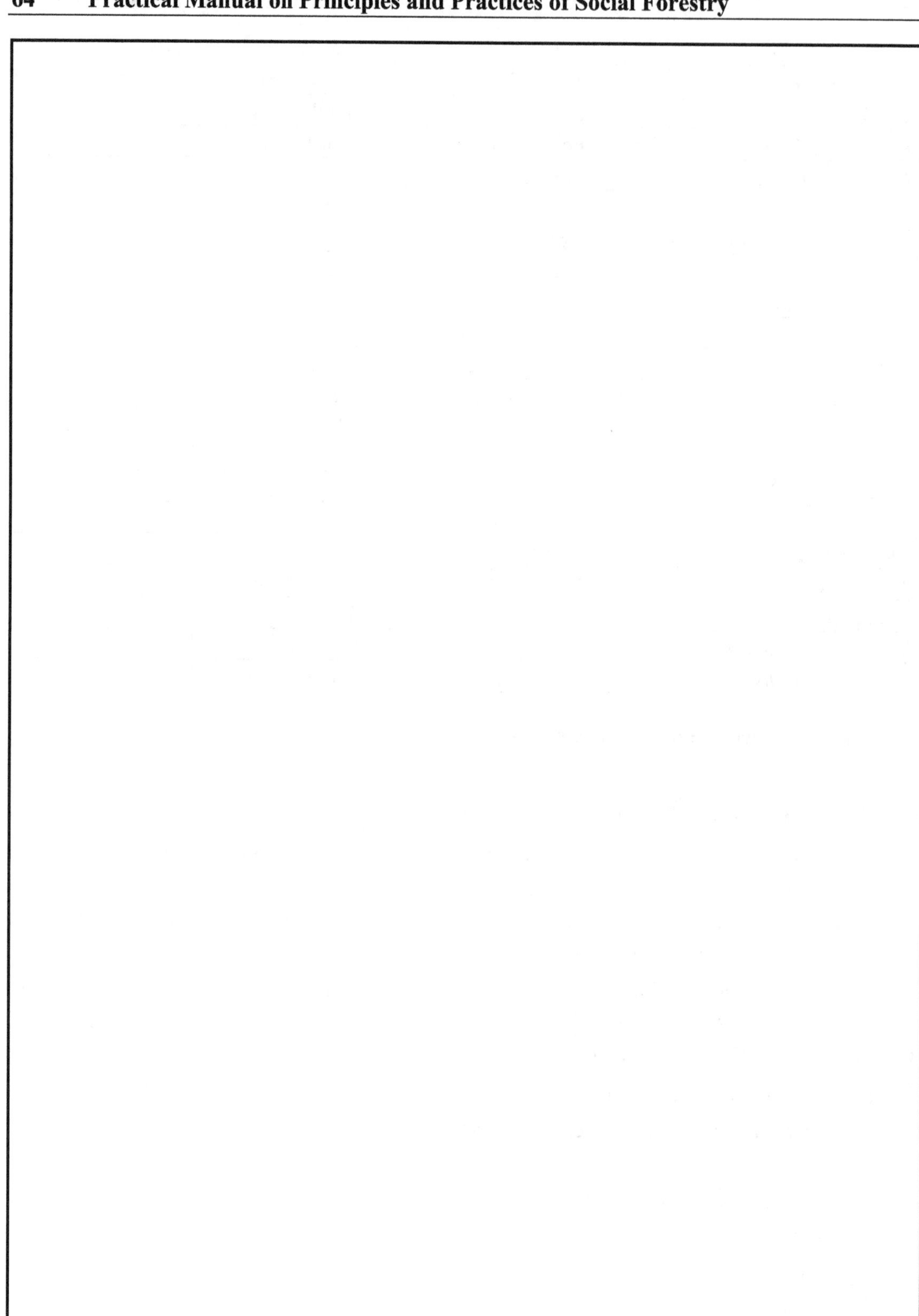

(c) Guidelines for seed collection

1. As far as possible seeds should be collected from seed production areas of stands, elite or plus trees, seed orchards and certified trees.

2. Seeds should be collected from the best stands in the identified ecotype.

3. Seeds should be collected from the trees or stands which are middle aged to near maturity, dominant, vigorously growing, having good form, straight cylindrical boles and clean stems, good crown development and free from defects and other diseases and pests.

4. Seeds should be harvested only when they are fully ripe and as far as possible from the trees.

5. Trees growing in isolation, open and on agricultural lands should be avoided.

6. It is desirable to collect seeds during good seed years, because during poor seed years, the quality of seed is very poor.

7. Species having wide altitudinal range of dispersal require high care. Such seeds should be collected from proper zones, so that seeds collected from the high zone are not used in the lower zone and vice-versa. Similarly seeds collected from the dry zone should never be used for sowing in the moist zone and vice-versa.

(d) Methods of collection

There are two broad methods of collection of seeds.

(i)   Collection from the forest floor which are either natural fallen or by shaking of trees.

(ii) Direct collection from the standing trees or by lopping the branches.

(i)   Collection from the ground

Only those species that have large and heavy seeds or bulky fruits and which fall unbroken, do not deteriorate soon may be collected from the ground. Only freshly fallen seeds are to be collected and for this purpose day to day cleaning and sweeping the floor before collection is necessary. Insect damaged and diseased seeds should be rejected. Only trees which are present on a flat and dry ground should be selected for ground collection of seed. Seeds of *Sal, Teak, Oak, Gmelina, Artocarpus chaplasha, Anthocephalus cadamba* etc. are collected by this method.

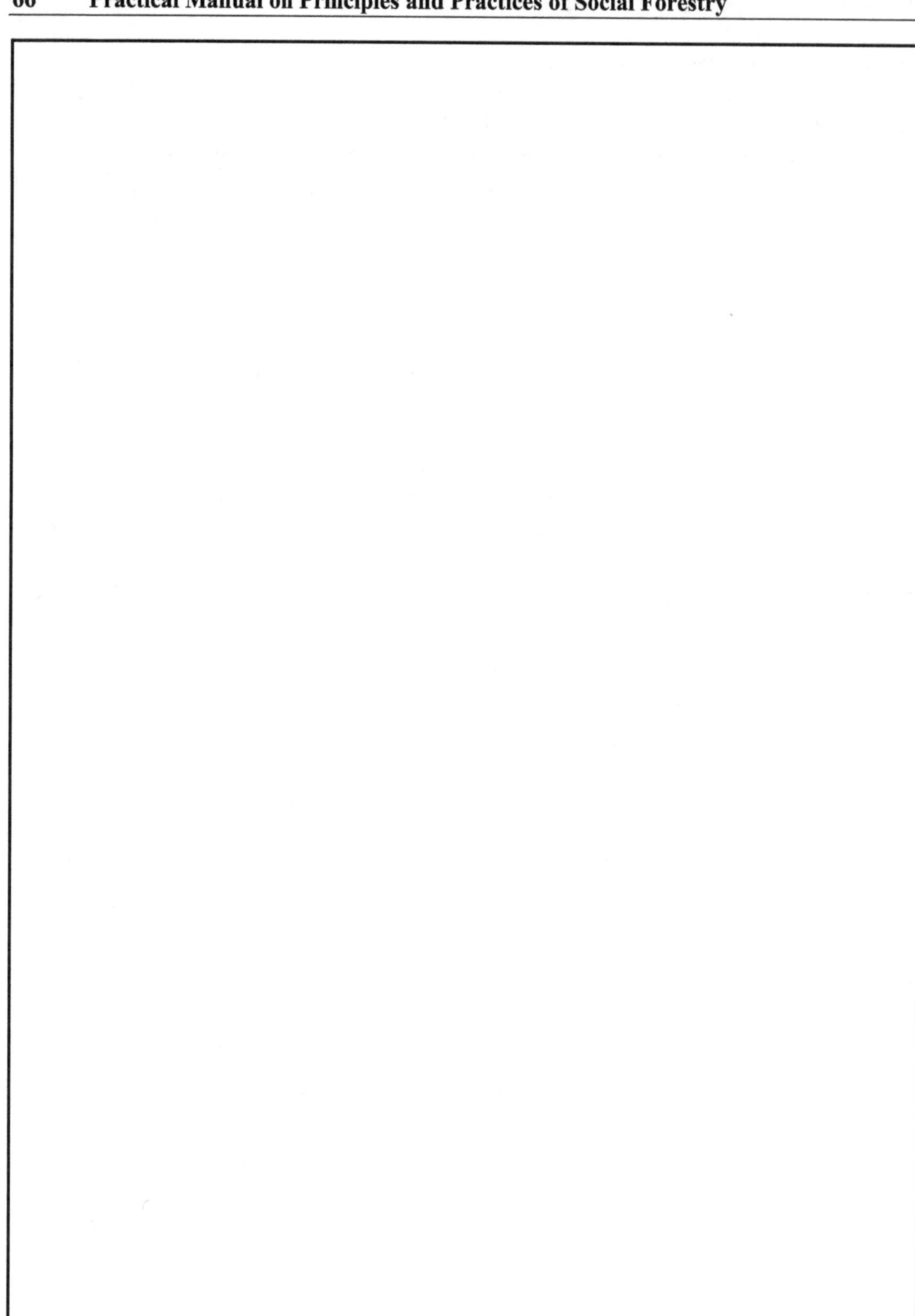

(ii) Direct collection

Collection from standing trees is usually more time consuming than from the ground or felled trees. Often collection can be made from tops of trees felled during lopping operations provided they are felled at proper stage of seed maturity. Various types of tools have been adopted or developed to facilitate cutting of cones, fruits, and twigs from standing trees. Ladders and climbing irons are also used. In India mazdoors are engaged to climb up the trees and of the branches of good seed trees. Here care must be taken that least damage is done to the fruit bearing trees. The boles of the trees should not be spoiled by hacking out footholds and lopping should be confined to smaller twigs and their branches. Although our seed collectors are efficient at climbing, climbing apparatus is generally not available. By this method, seeds of *Acacia catechu, Dalbergia sissoo, Albizzia* spp, *Holoptian, Ailanthus conifers* etc. can be collected.

When many seeds are borne within large fruit or cone as in case of conifers, the entire fruit is usually collected. Small seeds as in birch (*Betula* spp) and poplars may conveniently be gathered by clipping off small twigs and branches with mature fruits attached.

Seed collection equipment

In the west, seed collection is done in these days mechanically. Tree shakers and tree ladders which can be extended to the top of the trees, branch to branch are already in use by which easy collection is possible. Hydraulic cone pickers are also available. Dewingers specially designed for oak seeds have been in use in Sweden.

The most commonly used equipment in India are:

1. Tarpaulin material

The tarpaulin must be strong enough to resist handling, puncturing or tearing and should minimize condensation. The best material is heavy weight water proof canvas. Plastic coated cloth tarpaulins are not recommended because they tend to promote condensation. Polythene sheet also should not be used as they are easily ruptured. The size should be smaller which three (3) persons can easily manage. A size of at least 5 × 7 m to 5 × 10 m is suitable.

2. Nets

In the United Kingdom, a triangular shaped net used for seed collection from trees bearing small sized fruits has been built. The net is suspended by special ropes and notch blocks from the crown of the tree. The operator can climb up the net safely and is able to collect small fruits, which is otherwise not possible.

3. Portable ladders

Ladders are of varying lengths, of which length can be extended or reduced. Sailors ropes are the generally used with the ladders for collecting seeds in the countryside. These are knitted of long fibrous material and can be extended by making knots. The other one is one-legged ladder. They are suitable for uneven terrain and branchy trees. The one legged ladder consists of one pole (hence one legged), with alternate short rungs on both sides. One-legged ladder can be attached to the trunk of the tree. Now-a-days the light weight aluminum scaling ladders are used for climbing tall trees. It consists of sections usually of 2-4 m length. The sections are designed to fit into one another and are attached by a chain to the trunk of the tree, which the operator grasp as he climbs up.

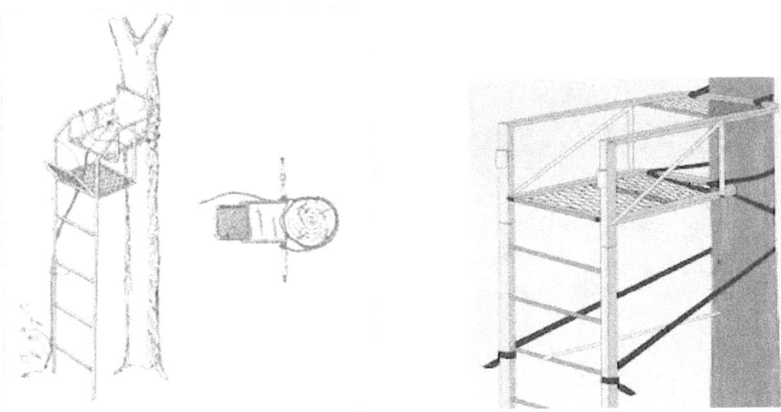

4. Tree bicycle

Tree bicycle are very easy to use after a little practice. They are used to collect seed or pollen on trees having smooth barks. It consists of two pedals supported by a vertical arm. The pedal is fastened by special straps steel band forming a circle of adjustable diameter around the stem of the tree to be climbed. The pedals are 5-8 cm greater than the diameter of the tree. In climbing, the operator puts all his weight on one foot peddle, then raises the other foot up to meet it. Between the two operations, he must adjust his safety belt as high as possible. When the operator reaches the crown he can take off the pedal and fasten them to trunk for freedom of movement.

A German investor has patented a bicycle that can climb trees and can also be used as a paddleboat

5. Climbing irons

It is an old method of climbing up the trees. The forged iron is fastened to the foot gear with two sturdy leather straps. The foot gear must be solid and firmly attached to the climbers foot and leg. The iron ends in a fixed spur of varying length. One of the belt has a shoe hook which does not extend past the sole of the boot. This enables the climber to walk on the ground without difficulty. Iron can damage the tree bark especially smooth bark trees and should be used with precaution.

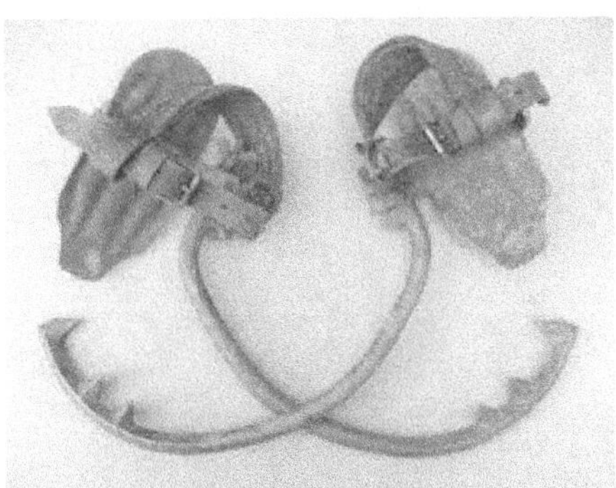 

6. Seed Extraction

Seed extraction is separating the seed from the fruit, cone, husk, stones or any other enveloping material. Species in which the whole or a part of the fruit is sown i.e., teak, walnut, ash, and dry fruited. *Terminalia* spp. need no extraction as they are ready for storage after drying and cleaning. However when the seed is the only part sown as in most of the conifers, dry fruits and a majority of the leguminous trees, seeds have to be extracted before they are stored.

From the viewpoint of extraction, tree seeds can be categorized into following groups.

1. *Fleshy fruits*: In this category comes the aggregate fruits berries, drupes and pomes. As a general rule, the pulpy portion should be removed as soon as possible, if the seed is to be sown immediately after collection as in case of *Artocarpus, Michelia* etc., failure to do so seriously lowers the germination percentage as in case of *Melia azaderach, Azadirachta indica* etc. In all cases the seeds can easily be extracted by macerating the fruit in water and allowing the fleshy pericarp to float away.

2. *Dry fruits*: Dry fruits with seed surrounded by tightly adhering pericarp are of three types.

    (a) Achenes which may be free of appendages or retaining the feathery styles i.e., *Elematis, Alstonia, Bombax ceiba*, poplars, and willows.

    (b) Nuts: i.e. oaks, chestnut, walnut, cashewnut etc.

    (c) Winged fruits or samaras i.e. *Fraxinus,* Acer, Dipterocarpus, Hopea, Shorea, Moringa, *Ailanthus excelsa, Dalbergia sissoo* etc.

Seeds of this category are rarely extracted from the fruit. For practical purposes, the entire fruit is seed, although sometimes the styles of some achenes and wings of samaras are removed to reduce bulk and facilitate handling. In case of legume pods, the seed is extracted very easily by beating up or simple drying in sun to get separation of seeds.

In case of conifers, the complete extraction of seed is required before sowing. The usual method of extraction consists of spreading the ripe cone in the sun on clean platform or in trays until they open up. They can then be shaken or beaten up to separate the seed. In case of *Bombax ceiba,* the opened fruits are kept in a gunny bag or bamboo basket and churned with a wooden stick till the seed is separated and collected from the bottom of the bag.

3. ***True-seeds***: They are readily extractable from dry fruits or pods such as *Acacia catechu, A. modesta, Parkinsonia aculeate* and capsules *(Salix, populus* spp). Extraction of seed from such fruits usually involves drying by solar or artificial heat followed by threshing or shaking on a sieve with a mesh larger than the seed and smaller than the fruit.

4. ***Dehiscent pods***: Usually open automatically when dried or broken up easily. Some of them like Buxa, and Bauhinia dehiscent lengthy and these together with those containing light winged seeds eg: Jacaranda, Spathodea should be covered over during drying to avoid loosing the seed.

5. ***Indehiscent hard pods, capsules or drupe***: Are often difficult to deal with *Pterocarpus* and *Tectona grandis* can be separated during pre-sowing treatments. Teak seed from seed balls can be extracted by splitting the fruit into two halves. Seeds of many species like *Cassia nodosa, C. fistula, C. javanica* and similar species can be broken up by pounding them in a mortar.

6. Seed of few species like *Terminalia* spp. *Acacia catechu, Albizzia lebbek* is attacked by insects and collection done immediately after the seed is sufficiently ripe for extraction.

7. Most of the seed after extraction require drying, although seeds of some species loose their viability by drying. Seeds should be extracted and air-dried in the shortest possible time in such a way that there is least damage to the seed.

   (i) ***Artificial***: Drying of fruits in heated / solar kilns has been used in some countries. Artificial heating in kilns permits control of air, moisture and temperature with a continuous process. This practice is useful in drying of cones.

   (ii) ***Natural Drying***: Natural drying process involves less risk of lowering seed quality but takes longer time. This process cannot be used in very wet or rainy days. The following are the methods of natural drying.

        (a) ***Drying under cover***: In this method fruits are kept in well-ventilated rooms, spread thinly, stirred regularly and placed on trays with a wire mesh bottom to allow all round circulation of air. Though it is a slow process but a safest method for delicate seeds which cannot withstand rapid drying eg: Deodar, Oak, Abies, Dipterocarpus, Hopea etc.

        (b) ***Sun drying***: This method is well suited to drying of cones and fruits of species which can withstand high temperature.

Final drying: Seeds of many species require a final drying before storage to reduce the moisture content to 10-12 % because seed with higher moisture content before storage are likely to be damaged by micro organisms.

Moist extraction

There are certain kind of seeds such as *Shorea robusta, Azadirachta indica, Syzygium cuminii,* oaks and many Lauraceae family which loose viability by drying, such seeds have to be extracted and stored in moist conditions.

SEED STORAGE

Seed storage is the preservation of viable seeds from time of collection until they are required for sowing.

Necessity of seed storage:

1. Trees do not bear seeds regularly year after year, the time of collection of seeds does not coincide with nursery sowing, therefore to conserve their viability and germinative capacity, seeds have to be stored for making reserves for poor seed years.

2. Seeds required after ripening.

3. Seeds are to be transported to long distances.

4. When trees do not bear seeds annually such as for Deodar, Bamboo, Spruce, Chirpine etc.

5. To protect the seeds from rodents, birds and insects.

Factors effecting longevity of seeds during storage

1. **Seed condition:** Even under ideal storage conditions seeds will loose viability if it is defective from the start.

2. **Seed maturity:** Fully ripened seeds retain viability longer than seeds collected when immature, as certain bio-chemical compounds necessary for seed viability and also the dormancy inducting compounds are synthesized during final stage of seed ripening.

3. **Parental and annual effects:** The percentage of sound seeds in high yielding mother trees and in a good crop year is usually higher compared to poor yields and in a poor year.

4. **Freedom from mechanical damage:** Seeds damaged mechanically during extraction, clearing etc. rapidly loose viability.

5. **Temperature, moisture and oxygen:** The respiratory process depends on moisture content in the seed. The purpose of lowering temperature, moisture and oxygen during storage is to lower the respiratory activity to ensure that the seed maintains its germinative capacity for a long time.

6. **Freedom from fungi and insects:** For this purpose fumigation with methyl bromide, carbon bisulphide is used commonly to kill insects and seed treatment with fungicides to control fungal growth to store the seed in good viable condition is necessary.

Storage and viability

1. Species which retain their viability for short period when urgently needed to be stored for a few days more, should be stored under shade to prevent injury due to desiccation and giving them sprinkling of water from time to time. Sal, Dipterocarpus and many seeds of Myrtaceae and Lauraceae can be stored in this way for 2-3 weeks.

2. Seeds of temperate species which ripen in autumn and germinate in the following spring, can be stored best under low temperature. Cold storage has been a very useful method for these species.

3.   Most of the species in the plains produce seeds in winter or summer and germinate in the following rainy season. Thus they can bear the varying temperatures and moisture conditions. Such seeds need to be stored in dry conditions in sealed containers.

Storage containers: Storage containers suggested for Indian tree seed are gunny bags, sealed tins, perforated polythene bags, cloth bags etc.

Normal efficient storage period for different species

S. No	Normal efficient storage period not exceeding	Species
1.	Two weeks	*Dipterocarpus* spp, *Shorea* spp, *Hopea* spp, *Michelia champaca*, *Azadirachta indica* etc.
2.	Three months	*Chloroxylon* spp, *Mangifera indica*, *Pterocarpus* spp
3.	Six months	*Abies* spp, *Alnus* spp, *Betula* spp, *Eucalyptus globules*, *Jugalans regia* etc.
4.	One year (in sacks)	*Tectona grandis*, *Terminalia tomentosa*, *Bambusa bamboos* etc.
5.	One year (closed tins)	*Acacia catechu*, *Anacardium occidentale*, *Dalbergia latifolia*, *Eucalyptus* spp, *Pinus* spp, *Bombax ceiba*, *Dendrocalmus strictus*, etc.,
6.	Two years or more	*Acacia nilotica*, *Albizzia lebbek*, *Albizzia procera*, *Dalbergia sissoo*, *Prosopis juliflora*, *Cassia fistula*, *Pinus roxburghii*

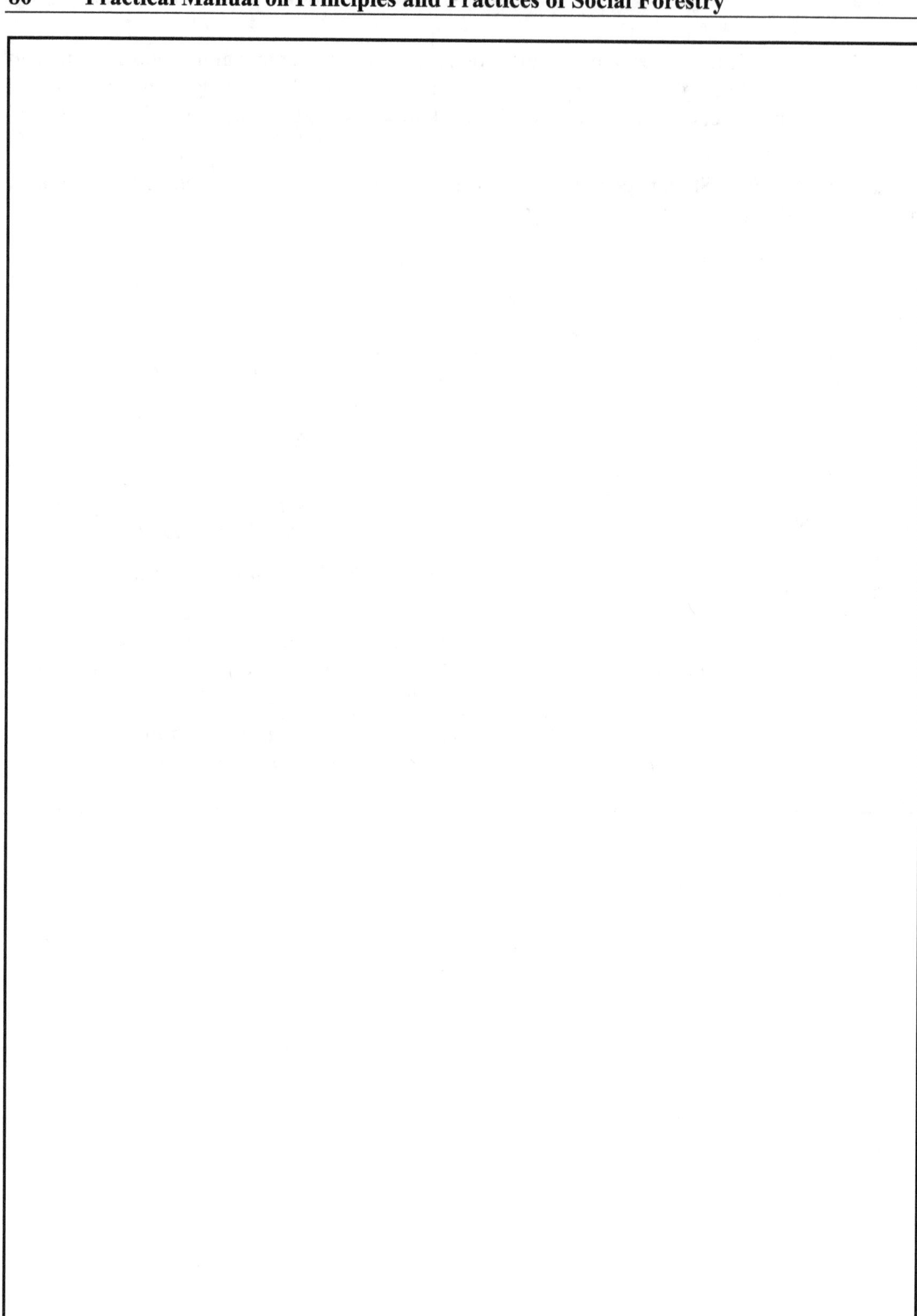

TESTS FOR SEED VIABILITY AND GERMINATION

Objectives:

1. ____

2. ____

3. ____

Seeds are capable of withstanding adverse environmental conditions for long periods and they usually germinate when conditions become suitable. Therefore, the seeds have a life span of their germination, more correctly the time when they begin to loose their capacity to germinate. This period is called viability and it varies from species to species.

Determination of seed viability

Viability, sometimes called seed soundness is the percent of seeds capable of germinating, when exposed to the most favourable conditions. This may be determined either directly by germination tests, indirectly by such physical measures by cutting tests, the growth of excised embryos, floatation, size, colour and such bio-chemical methods as staining of embryos and measurements of enzymes activity.

(A)   Physical Tests

1. *Cutting test*: It is the most inexpensive test for testing viability. The cutting test reveals the emptiness, decay, insect attack etc. As the seeds of *Azadirachta indica* loose viability quickly cutting test is performed to know the soundness of seeds, seeds having yellow or brown cotyledons are termed unsound and unfit for germination.

   To carryout this test, a representative sample of seed is taken and spread out on a table. Each seed is then cut open or broken with gentle strokes of a hammer, the cross sections examined with a hand lens or reading glass and percent of viable seeds determined visually. Seeds which in cross-section appear firm, plump and of good colour are considered viable. Mouldy, decayed, shriveled, rancid smelling and those lacking embryo and endosperm (Bling), abortive, etc are grouped as non-viable.

2. *Floating test*: Since the empty seeds of many species float in water, particularly in heavy seeded species such as oaks, beach, and the sound ones sink, this method is sometimes used to measure the seed quality. If the seeds are not freshly collected and have become rather dry, many sound ones may also float. For seeds not as heavy as water, light liquids and for others, alcohols have been used to separate viable from empty seeds.

3. *Size*: The number of sound seeds per kg is occasionally used as an index of seed quality. Large seeds from given climatic zone are usually are preferable to small seeds.

4. *Colour*: Color changes in fruits provide a single and reliable criteria for judging seed maturity. In *Ficus benjamina* seed collected at the time of brownish yellow colour of fruit gives better germination results than at other stages. Good seeds of conifers, have bright clean coats, seeds fallen on the ground change their colour quickly as they are stained due to moulds or fungal infection seed stored at low temperature and high humidity will develop spots of moulds on the surface of the seed coats. In case of *Dalbergia sissoo,* the black seeds are less viable than brown seeds.

(B)   Bio-Chemical Methods

(a) *Staining seed embryos*: This method is based on the principle that only living tissues stain certain dyes, other only by dead material. Indigo carmine for instance, will stain dead tissue, salts of selenium on the other hand will stain living embryos to red colour. Potassium iodide chemical stains living tissue blue.

The recent and widely used test of viability is the 2, 3, 5 triphenyl tetrazolium chloride staining method (TTZ). The chemical is colourless in the normal (oxidized) form but when reduced by living tissues (as by enzymes Dehydrogenase present universally in living seeds), it attains intense red or orange colour. Thus the living tissues in the seed stain, while dead tissues remain colourless. The test can be performed as follows.

Procedure: Moisten the seeds if they are dry by keeping them in moistened blotter or paper towel. Then gradually bring it into water and keep overnight at 30^0 C for 18 to 24 hours. The seeds should not show any sign of sprouting. Hard coated seeds are scarified before placing in water. The time of soaking can be reduced or increased according to the type of seeds. Remove the seed coated carefully and expose the embryo or divide the seed into two parts through the embryo. Small seeded legumes can be placed directly in the solution. An aqueous solution of 0.25 to 1 % TTZ solution of pH 6-7 is used. To prepare a 1% solution 1 gram of tetrazolium powder is dissolved in 100 ml distilled water. The solution must be stored in the dark to prevent deterioration from light. Used solution is discarded after each test. Take the solution in a petri-dish and place the embryos in it and keep it in darkness at 30^0 C for 1-18 hours according to the type used for testing. After keeping the embryos for required time, remove them from the solution and wash off the excess solution in cold water. Washing in water should not cause any harm to seeds. Then examine the embryos under a stereomicroscope for coloration. The viable seeds will develop a normal red colour and non-viable will not. However, it requires special training and standardization of techniques and interpretation among analysis. Temperature between 20^0 C and 45^0 C have no effect on accuracy of tetrazolium test, but staining proceeds faster at higher temperatures.

Interpretation of results: It is done as follows:

1.   Seed completely stained – germinated and vigorous.

2.   Minor stained areas on cotyledons – germinate and vigorous.

3.   Extreme tip of radicle unstained – germinable, least vigorous.

4.   More than extreme tip of radicle unstained – non-germinable.

5.   Unstained area at the juncture of cotyledon and radicle hypocotyl axis – non-germinable.

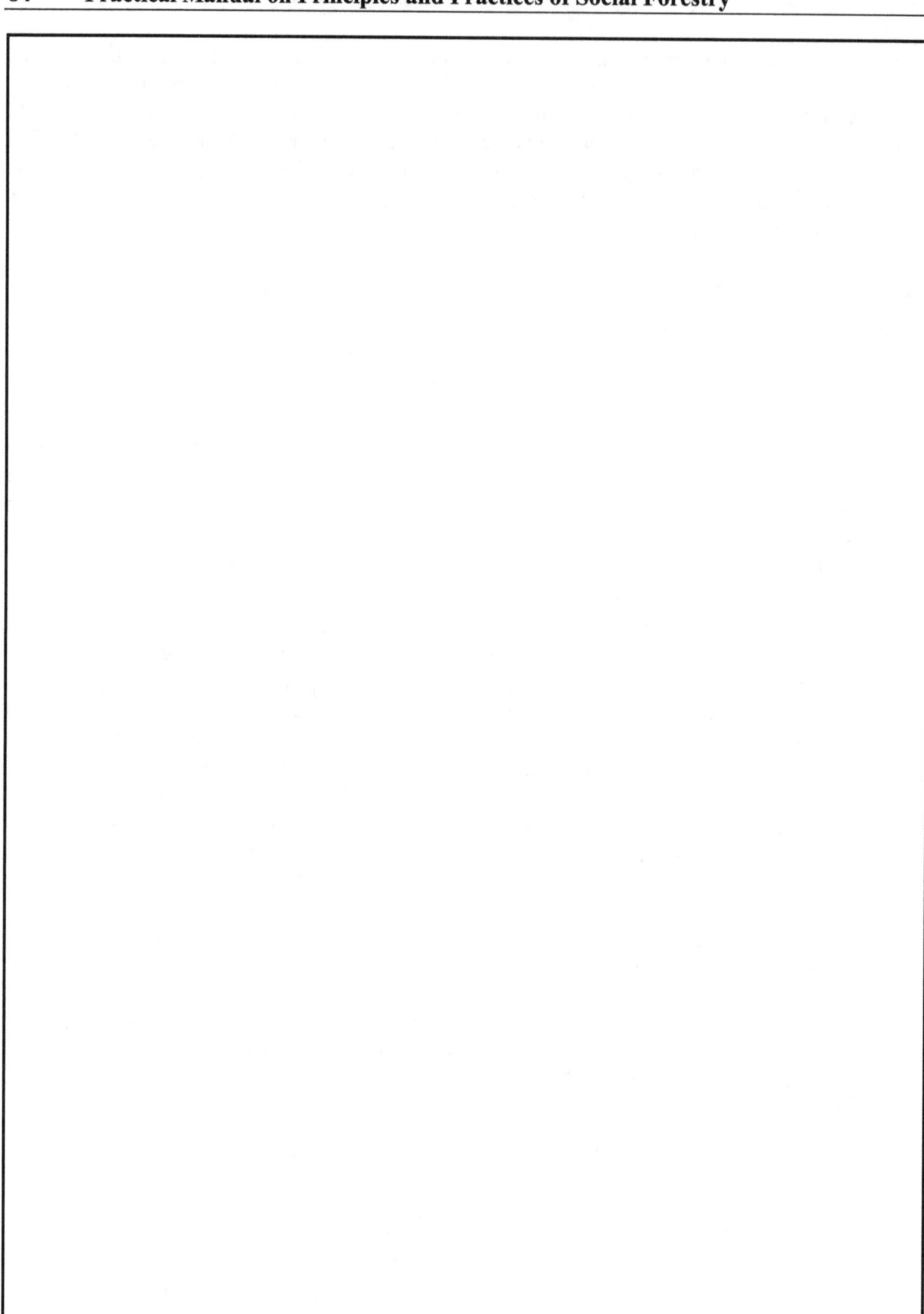

6.   Radicle hypocotyl axis bisected unstained – non-germinabale.

7.   More than half cotyledonary tissue unstained – non-germination.

8.   Radicle unstained – non-germinable.

9.   Seed stained off colour, grayish red, orange-red, or glassy or transparent red – non- germinable.

10.   Seed stained very light pink colour – non-germinable.

11.   Seed completely unstained – non-germinable.

As the structure of embryo varies in respect of different species, the system of valuation of the tests has to be determined separately for individual species. Use of this method has a series of problems, such as difficulties in staining some seeds, opening of some seeds for checking of colour and in some cases it is the difficulty with interpretation of different shades of colour.

(b) *Catalase activity*: Catalase activity of seeds can be a measure of seed viability. Fresh seeds are soaked in water at 54^0 C for an hour. Liberated oxygen is measured and thus quantitative determination is done. The catalase ratio is expressed as:

$$\text{Catalase} = \frac{\text{Catalase activity of soaked seeds}}{\text{Catalase activity of dry seeds}}$$

If the ratio is more than one, the seed lot is considered to be good. A catalase apparatus should be assembled having reaction chamber.

X-rays test: X-rays have also been used for testing the viability of seeds. In this test the seeds are first soaked in water for 16 hours at room temperature and then in a solution of barium chloride for 1 to 2 hours. While barium chloride penetrates dead tissues through process of diffusion, it cannot penetrate the living cells due to their semi permeable nature. The seeds are then washed for 1 to 5 minutes in order to remove barium chloride from the surface and then dried in a thermostat at 70^0 C for 2 hours. Finally, they are photographed with soft X-rays and exposure time of 2 seconds on 1-15 roentgen rays. An embryo is considered viable if it is free of impregnated and the endosperm impregnated not more than 25%. The main advantage of this technique is that the seed can be used even after the test. With this method, it is possible to separate the empty seeds from the filled seeds. It facilitates to study the development of embryo and endosperm in a seed to distinguish between the monoembryonic and polyembryonic seeds, to determine the number of seeds in an intact fruit, to observe the weathering damage on seed etc. By using barium chloride and organic contrast agents (X-rays contrast method), the utility of X-ray radiography has been extended for determination of germination capacity of some species of forest trees.

Decoating or excised embryo test: The technique consist of soaking the seed in water from 1- 4 days at room temperature until seeds are completely swollen. The water should be changed twice daily and after a few days of soaking, fully imbibed embryos are excised. Hard seed coats are cracked, mechanically scarified before soaking. The seed coats are removed with the help of scalpel or a needle. The decoated seeds are incubated in petri-dishes on moist filter appears or moist sterile cotton at temperature suitable for different species. Some of the decoated seeds germinate shortly and produce normal seedlings, while other exhibit spreading or greening and growth of one or more of the cotyledons. The embryos are examined daily up to a maximum of

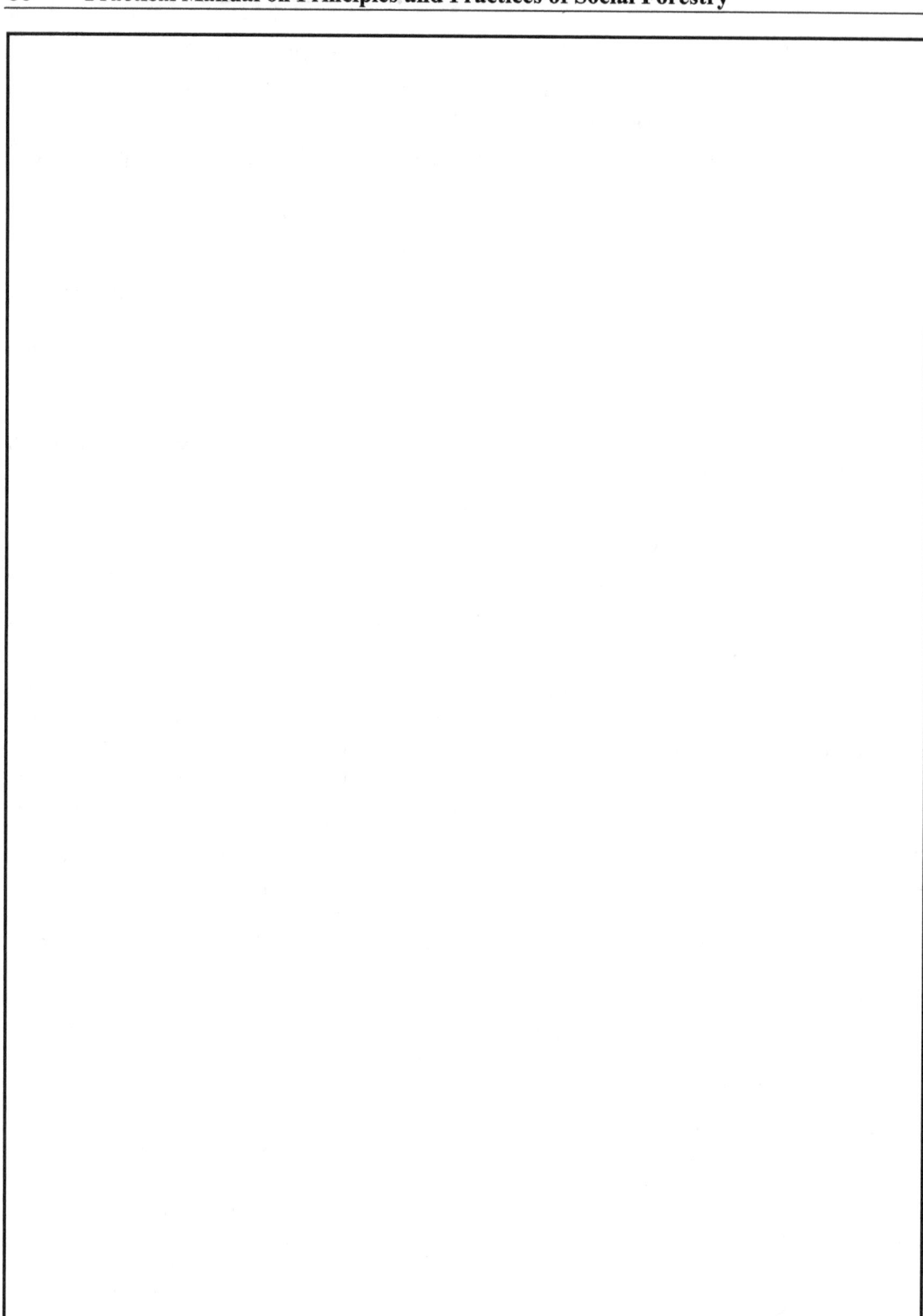

14 days. Weak or dead seeds deteriorate or exhibit brown or black colouration. The chief advantage of this method lies in the fact that it permits the determination of the viability of seeds within a few days or weeks otherwise requiring several months for germination in soil or other media.

Interpretation of results:

(a) The following categories are considered viable.

　　1. Germinating embryos.

　　2. Embryos with one or more cotyledons exhibiting growth or greening.

　　3. Embryos remaining firm, lightly enlarged and either white or yellow according to species.

(b) The following categories shall be considered non-viable.

　　1. Embryos which rapidly (develop severe mold) deteriorate and decay.

　　2. Degenerated embryos.

　　3. Embryos exhibiting extreme brown or black discoloration, an off-gray colour or white, watery appearance.

　　4. Dead or embryoless seed detected during preparation and seeds with deformed embryos.

At present various types of germinators are available for seed testing. One such is the thermo statistically controlled germinator having temperature range from 5^0 C to 56^0 C (Copenhagen type). Though a large number of devices are available yet the following are used for medium sized and large seeds for testing.

　　1. For medium sized dry seeds, a porous fine clay plate with 50 or 100 depressions standing in a dish or water, fitted with a glass cover to maintain a most atmosphere is used. One seed is placed in each depression and as it germinates, it is removed, keeping a count of such removals.

　　2. For large seeds, a shallow wooden box or an ordinary flower pot filled with clean sand, saw dust or light-soil, is used. Vermiculite mica is also used as a medium as it has many advantages.

Substrate for germination tests

Various types of substrata are used in germination tests, for example germination blotters, filter paper, toweling etc., free of toxic chemicals. The choice of media however depends upon the apparatus, the species and the working conditions in the laboratory. Two main types of media are used for germination tests viz. natural and artificial.

　　(a) *Natural media*: Fine non-alkaline sand, acid peat or shifted sphagnum mass are the preferred natural media. Non-alkaline sand contains only a few micro organism and hence can easily be sterilized. Seeds may be sown in a uniform layer of sand covered to a depth of 1-3 cm or they may be pressed into the sand. Mostly the seeds are germinated in sand moisture to 50-60% over its water holding capacity. Peat and sphagnum mass are practically free from the fungi that caused damping off. Garden soils contain large amounts of organic matter and is likely to be teaming with micro organisms such as those causing damping off.

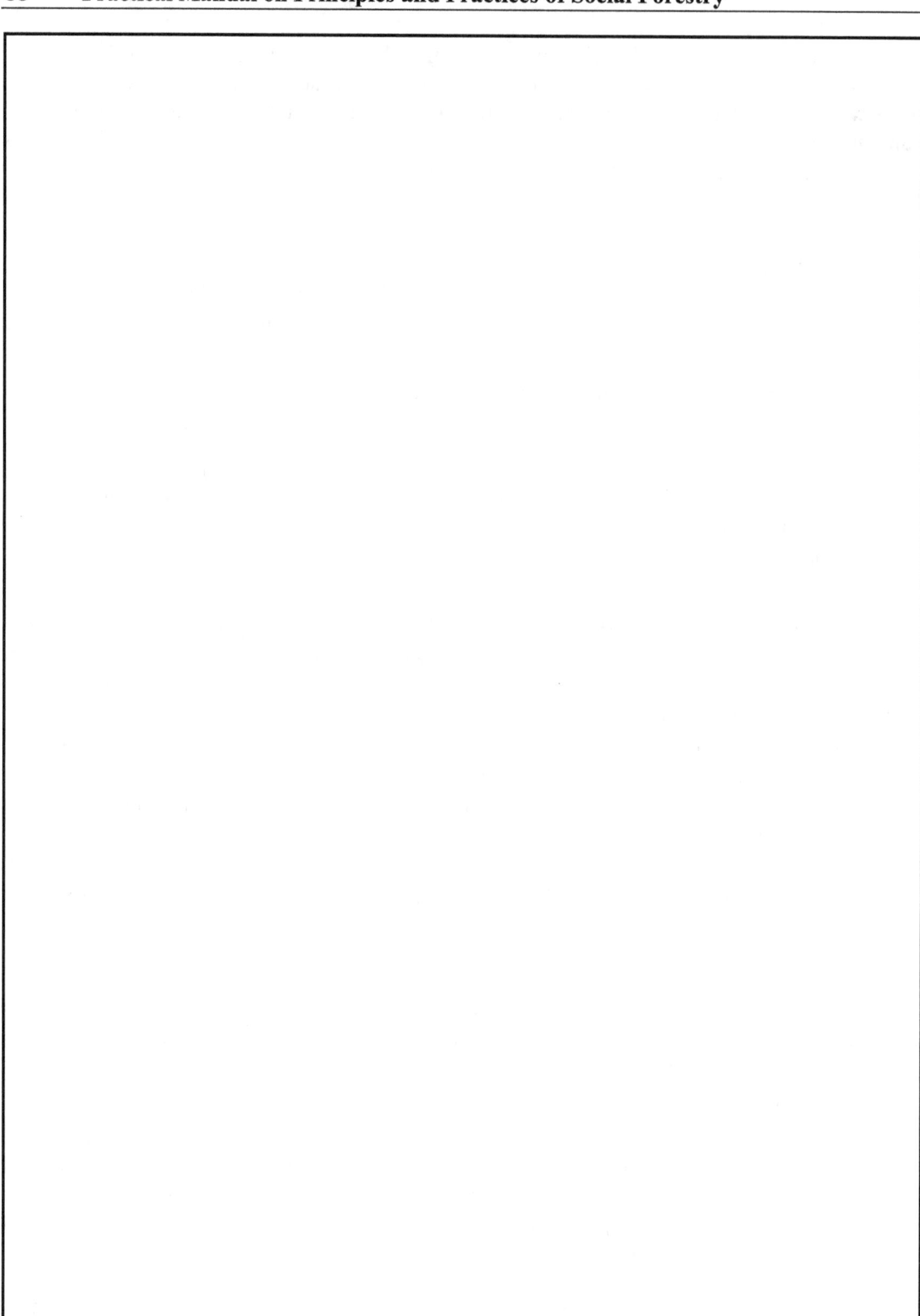

The chief advantage of using natural substrate like sand, peat or soils is that they give results close to those obtained in the nursery. Such media ensure a better balance between moisture and aeration, discharge the growth of mould on the surface of the seed and largely eliminate germinable but weak seeds which would have no chance in the nursery seed beds.

(b) *Artificial media*: Germinators and petri-dishes are commonly utilized, while using artificial media for germination tests. The artificial media generally used are porous plates, blotters or filter paper towels or agar. The main advantage of using these artificial substrata is that they require less room and time for germination and that because of smaller bulk, temperature and moisture conditions are more easily controlled in germinators.

Sterilization of seeds and germination media: Seeds and germination media need to be sterilized before placing the seed in the germinator. Seeds are surface sterilized in 1 % mercuric chloride and excess chemical is washed off in running tap water. Care must be taken to prevent water fracture of seeds. Sterilization of petri-dishes along with the medium or the germination boxes is done in autoclaves at 110-120° C for half an hour. At this temperature, the spores of most of the phytopathological organisms get killed.

Germination: After sterilization the seeds are properly spaced on the moistened medium and are incubated at temperatures suitable to each species, germination is best obtained at 30° C for most of the species.

Results: The results of viability and germination tests may be described as follows.

1. *Percentage of germination* : This is the percentage of seeds in a lot which have actually germinated by the end of the trial.

2. *Germinative potential* : This is the sum of germinated seeds and the remaining sound un-germinated seeds.

3. *Germinative capacity* : It is percentage by numbers of seeds in a given sample that actually germinate irrespective of time.

4. *Germination energy* : This indicates the percentage germinated within a given time for example 70% within 7 days or 90% within 14 days etc.

5. *Plant percent* : It is the number of seedlings resulting at the end of a determinate time in nursery or in plantation. It is always lower than germinative capacity.

6. *Germination value* : Germination value (GV) is a product of peak value of germination (PV) mean daily germination (MDG) i.e., GV=MDG.

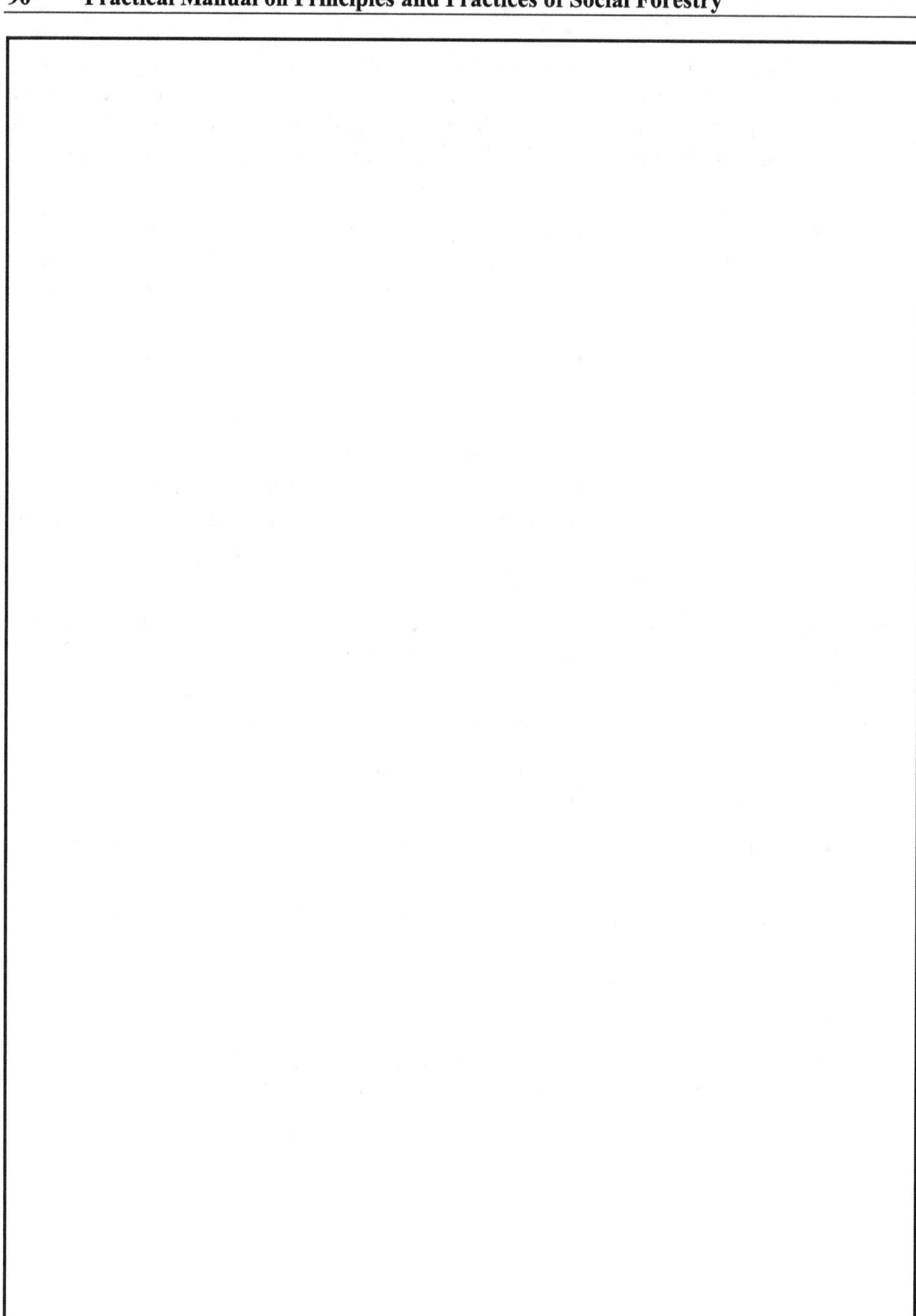

Species with their germinative capacity

S. No	Germinative capacity		Species
	Percentage	Qualitative	
1.	90 – 100	Very good	*Acacia catechu, Albizzia lebbek, Anacardium occidentale.*
			Bauhinia variegate, Biospyros melanoxylon, Shorea robusta.
2.	70 – 90	Good	*Azadirachta indica, Butea monosperma, Pinus roxbughii*
3.	50- 70	Average	*Bambusa, Dalbergia sissoo, Dendrocalmus strictus, Terminalia arjuna, Terminalia tomentosa*
4.	30 – 50	Poor	*Cedrus deodara, Tectona grandis, Bombax ceiba.*
5.	20 – 30	Very poor	*Abies pindrow, Cassia fistula, Casuarina equisetifolia*
6.	Less than 10	Extremely poor	*Alnus nitida, Anogeissus* spp

Average no. of seeds per kg, plant percent and germinative capacity of some important species

S. No	Species	Av. No. of Seeds / Kg	Plant Percent (%)	Germinative Capacity (%)
1.	*Acacia nilotica*	7,000-11,000	26	50
2.	*Abies pindrow*	27,200	6	13
3.	*Bombax ceiba*	21,430-38,500	30	75
4.	*Dalbergia sissoo*	53,000	45	90
5.	*Cedrus deodara*	7,000-9,000	58	65
6.	*Gmelina arborea*	2,500-2,600	30	85
7.	*Pinus roxburghi*	8,800-12,300	37	80
8.	*Shorea robusta*	575-1,000 (Fruits)	65	80
9.	*Terminalia tomentosa*	530	27	70
10.	*Tectona grandis*	1,850-3,100 (Fruits)	25	50

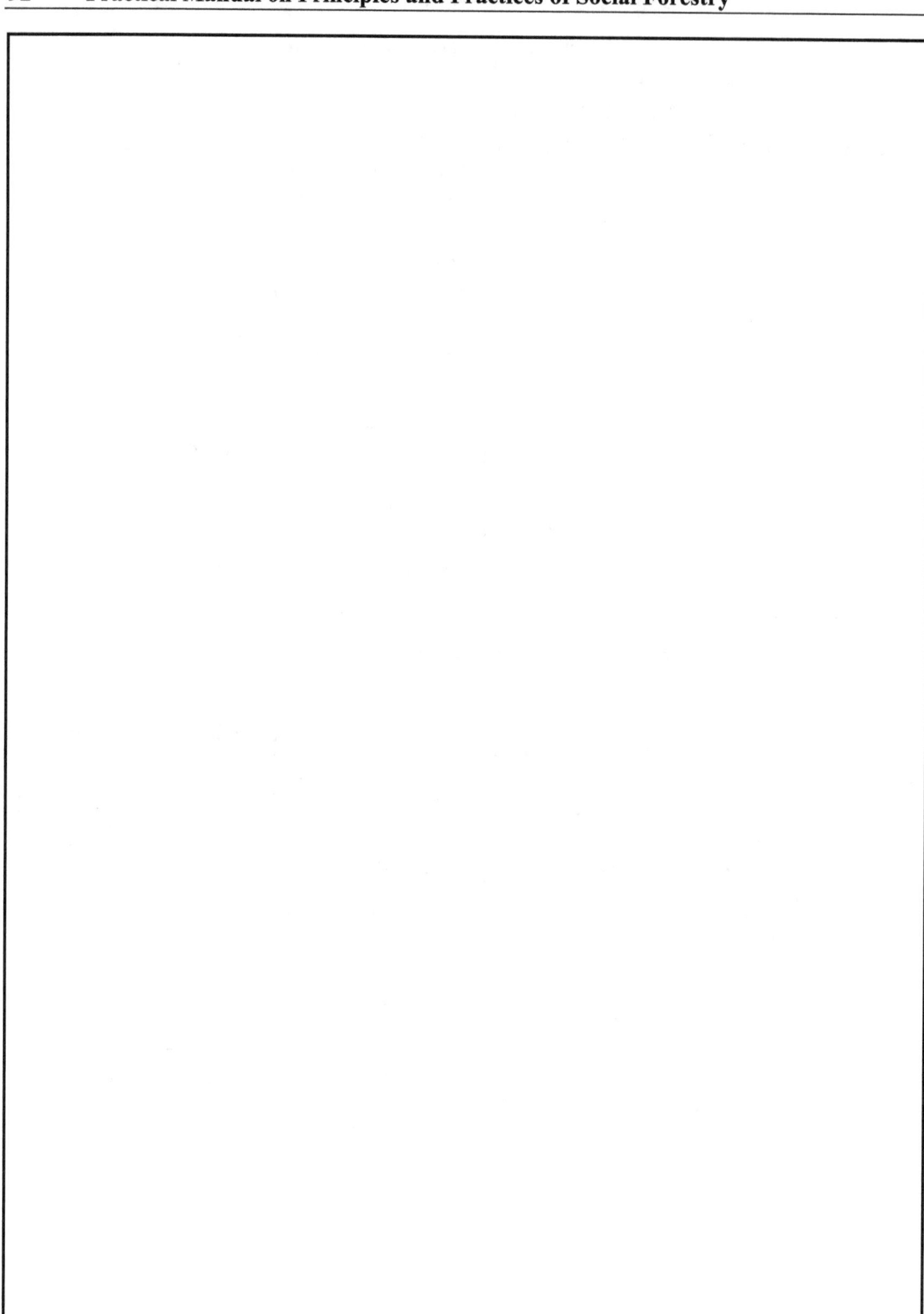

APPLICATION OF PRE-SOWING SEED TREATMENT TO TREE SEEDS

Objectives:

1. ____

2. ____

3. ____

For some seeds no pre-treatment is required as they are ready for sowing after collection, others however, require as they pass through dormancy during which the physiological maturity of embryo is attained. The nature of pre-treatment depends on the hardness of the seed coat. Seeds which have soft seed coats generally do not need any seed pre-treatment. However, it is advisable to put the seeds in water for 24 hours and dry them in shade before sowing. Seeds with hard coats can be subjected to a number of treatments which can be physical, chemical and biochemical method.

(a)   *Scarification*: Where germination is inhibited by mechanical resistance of the seed coat or impermeability of the coat to water or oxygen, dormancy may be broken by scarification. The term is used to define any treatment that renders the seed coat permeable to water and oxygen or weakens the seed coat so that embryo expansion is not physically retarded. In nature, seeds are scarified by microbial activity, forest fires or by passing through digestive tracts of animals or birds etc.

     Scarification may be accomplished in various ways, but may be roughly divided into two general categories.

1. *Mechanical Scarification* by scratching the seed coat by shaking the seeds with some abrasive material (e.g. Sand) or scratching or nicking the coat with a knife. The crack of scratch resulting for such treatment promote germination by decreasing the resistances of seed coat to water absorption and embryo expansion.

2. *Chemical Scarification* is an effective way of breaking dormancy resulting from the seed coat. Dipping seed into strong acids, such as acetone or alcohol can break dormancy. Even boiling water may be a successful treatment. As in mechanical scarification, chemical scarification breaks dormancy by weakening the seed coat.

Common methods of seed scarification

1. *Soaking in cold water*: Soaking in cold water for one to several days depending on the hardness of the seed coat. Water absorbed softens the seed coat, when soaking is required to be done for a long period, it is advisable to change the water regularly. After soaking treatment, the seed should be sown without loss of time as the drying seriously reduces the viability of seed. However in teak pre-treatment involves alternate soaking and drying.

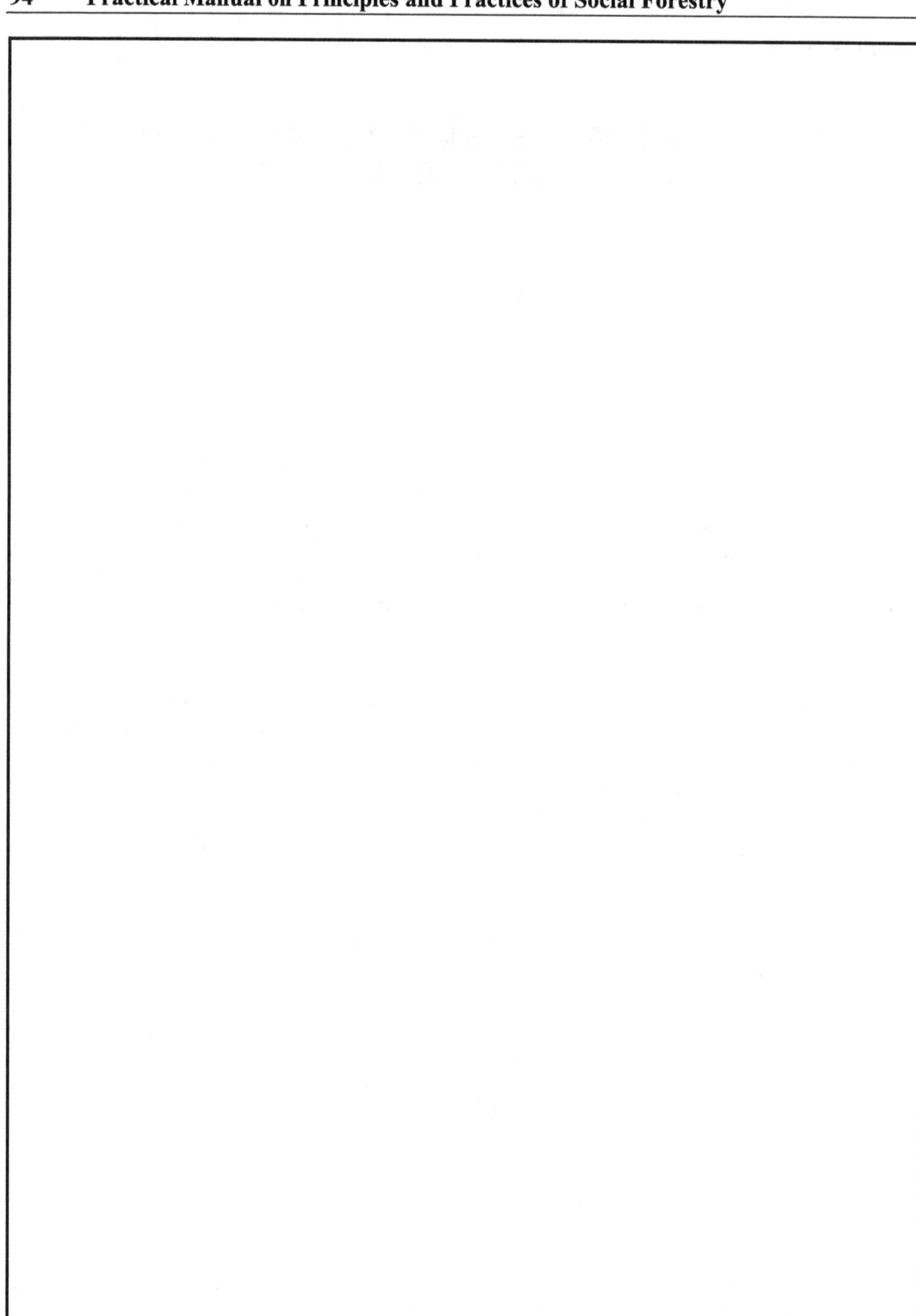

2. ***Soaking in hot water***: Seed of leguminous trees have extremely tough seed coat which still restrict germination for years together unless hot water treatment is applied. Seed is immersed two or three times its volume of boiling water and allowed to soak until it turns cold.

3. ***Acid treatment***: In this method the seeds are soaked in concentrated H_2SO_4 (specific gravity 1.7) for varying periods depending on the species. After soaking the seed must be immediately washed in clean water.

The soaking period for different species is given

S. No.	Species	Time in minutes
1.	*Acacia albida*	20
2.	*A. nilotica*	60-80
3.	*A. senegal*	40
4.	*Prosopis chilensis*	40
5.	*Parkia* spp	20

4. ***Other methods***: Seeds of Acacia, *Prosopis* and *Santalam* are sometimes fed to goats, later collected from their droppings hasten the germination. More chemicals such as potassium nitrate, copper oxide, zinc oxide, hydrogen peroxide and gibberillic acid have been reported to enhance germination in a number of species.

(b) ***Stratification***: This is storing of seeds in near freezing temperature even if no medium is used. The medium for stratification should be kept moist throughout the stratification period. The seeds are first soaked for one or two days and after draining excess moisture, the seeds are sealed in polythene bags for storage at 0^0 C to 2^0 C. Time required to break dormancy for different species is different.

I. Seed of some important species which require no pre-treatment

Abies pindrow	*Dalbergia sissoo*
Aegle marmelos	*Eucalyptus citriodora*
Abina cardifolia	*Eucalyptus globules*
Artocarpus heterophyllus	*Eucalyptus teriticornis*
Azadirachta indica	*Hardwickia binata*
Bauhinia purpurea	
Bauhinia variegata	*Lagerstromia parviflora*
Bombax ceiba	*Madhuca longifolia*
Butea monosperma	*Melia azaderach*
Pinus roxburghii	*Morus alba*
Sterculia urens	
Shorea robusta	
Syzygium cuminii	

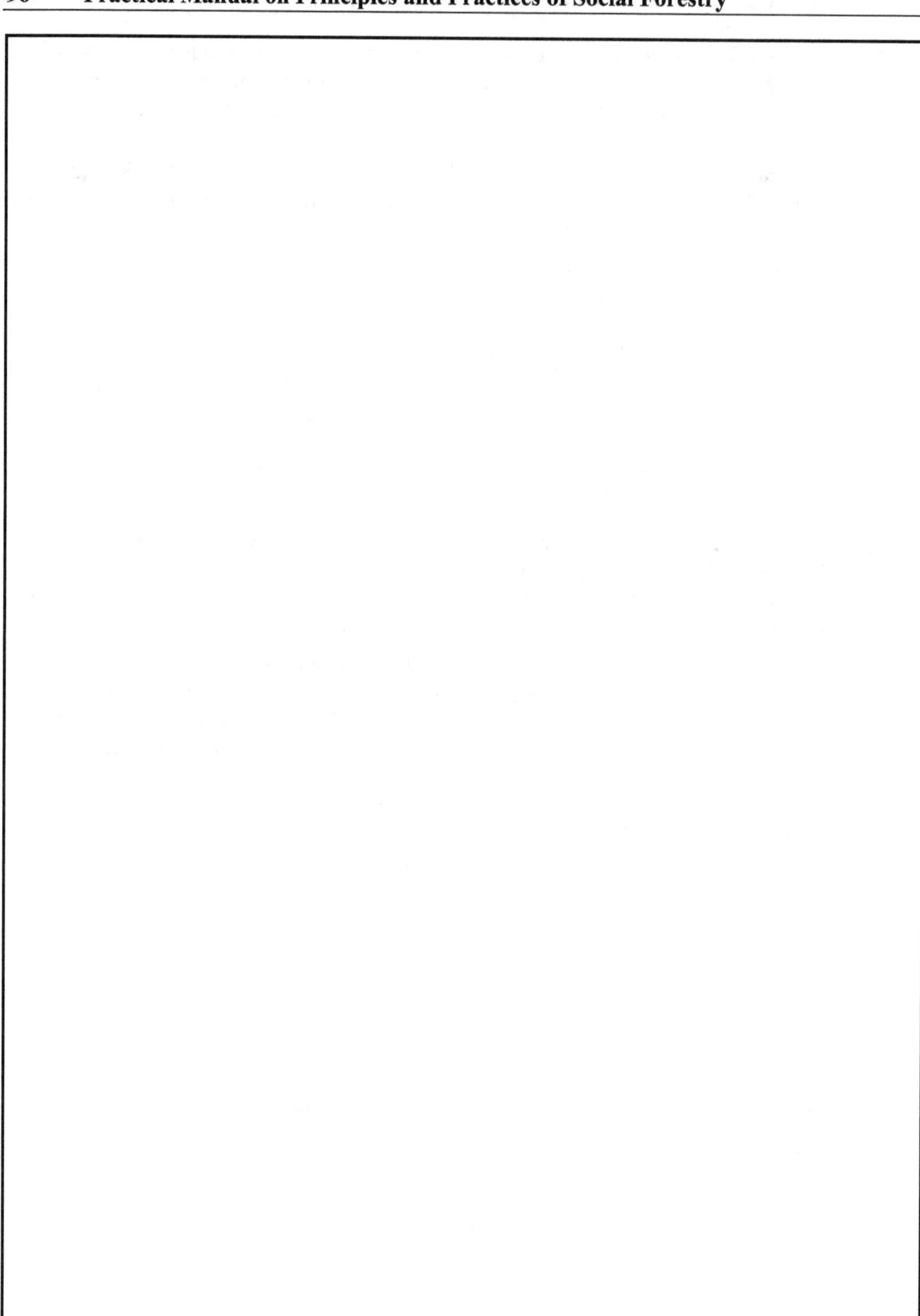

II.   Seed which require soaking in water

Anacardium occidentale	*Lagerstromia speciosa*
Anogeissus latifolia	*Mangifera indica*
Bambusa arundinacea	*Pithecalobium dulci*
Bauhinia racemosa	*Pongamia pinnata*
Casuarina equisetifolia	*Prosopis chilensis* (Soaking + Scarfication)
Cinnamomum camphora	*Pterocarpus marsupulm*
(in luke warm water)	(soak in camphor water)
Ficus spp (60^0 C for 10 minutes)	*Pterocarpus santalinus*
Gmelina arborea	*Santalum album*
Zizyphus mauritiana	*Tamarindus indica* (Soak in hot water)
Dendrocalmus strictus	*Terminalia alata*
	Terminalia arjuna

III.   Seeds which require scarification and hot water treatment

Acacia aurieuliformis	*Albizzia amara*
Acacia catechu	*Albizzia chinensis*
Acacia mearnsii	*Albizzia falcataria*
Acacia melenoxylon	*Albizzia lebbek*
Acacia modesta	*Albizzia odoratissima*
Acacia nilotica	*Albizzia procera*
Acacia senegal	*Cassia javanica*
Acacia tortilis	*Cassia siamea*
Artocarpus fraxinifoluis	*Emblica officinalis, Olea ferrugiana*
(Hot water, H_2SO_4 for about 6 Hours)	(Acid scarification)

IV.   Seeds which require stratification

Ailanthus altiossima	*Jugalans regia*
Alus nitida	*Micheilia champaca*
Cedrus deodara	*Morus indica*
Cuppressus torulosa	*Prunus nepalensis*
Fraxinus excelsa	*Quercus lineato*
Grevillea robusta	

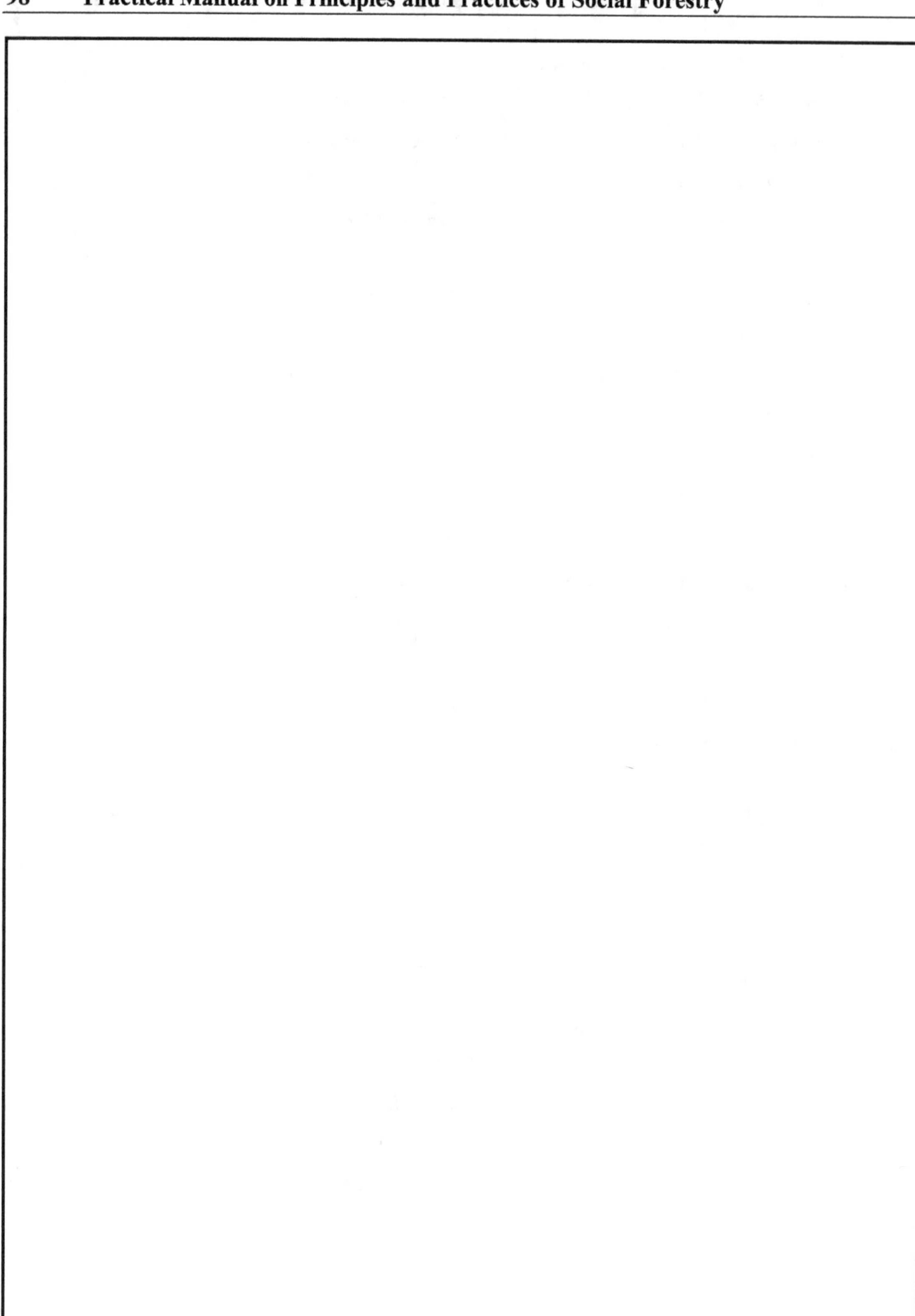

V.   Seed which require alternate wetting and drying

Diospyros melanoxylon

Tectona grandis

Terminalia bellerica

Terminalia chebula

VI.   Seed which require pre-germination

In some species, it is common practice to pre-germinate seeds in germination tray or nursery beds to ensure complete and uniform germination in the nursery beds. Germination is regulated by this method in the following species.

(a)   Pre-germination in trays/beds

Tectona grandis

Terminalia tomentosa

Pterocarpus spp

Eucalyptus spp

Dalbergia sissoo

(b)   Pre-germination in germination boxes

Adina cardifolia

Anthocephalus chinensis

Different Scarification methods

1. Mechanical scarification methods

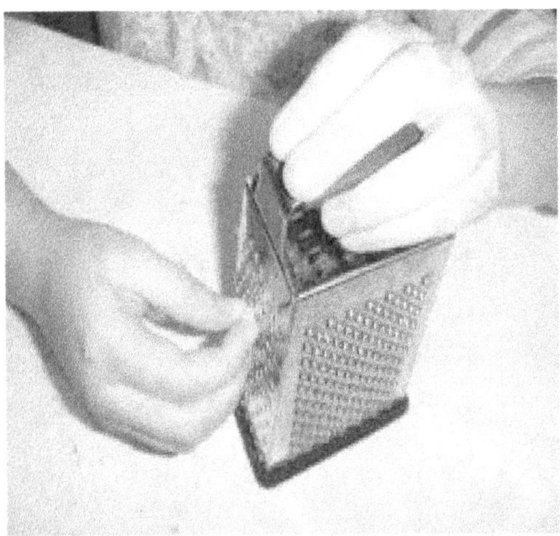

Rubbing against rough surface

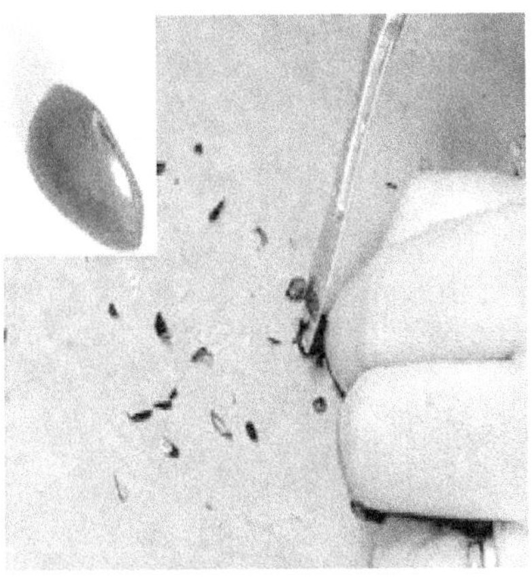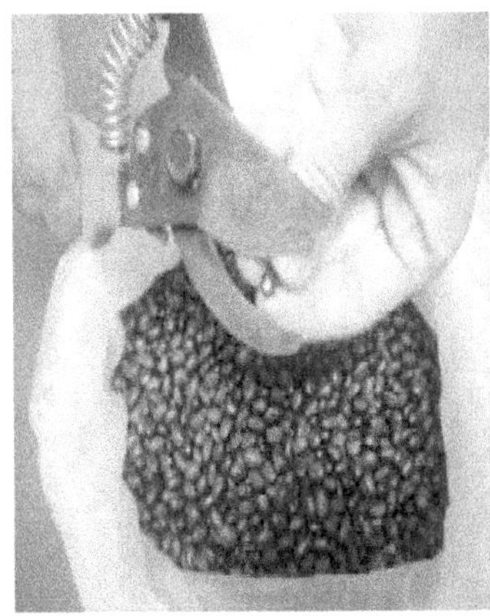

Making hole in the seed coat using different equipments

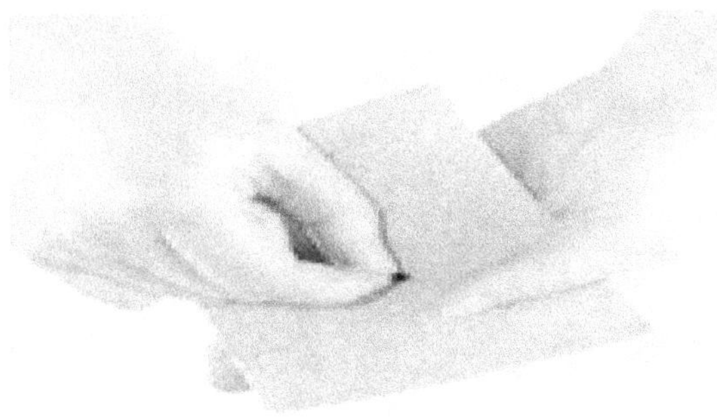

Rubbing against sand paper

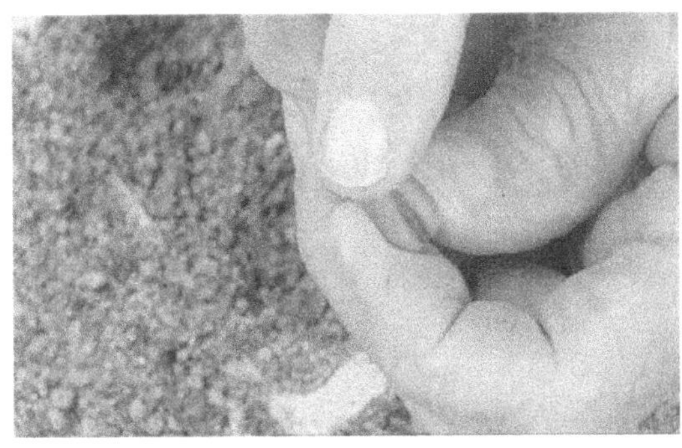

Making slight cut in seed coat

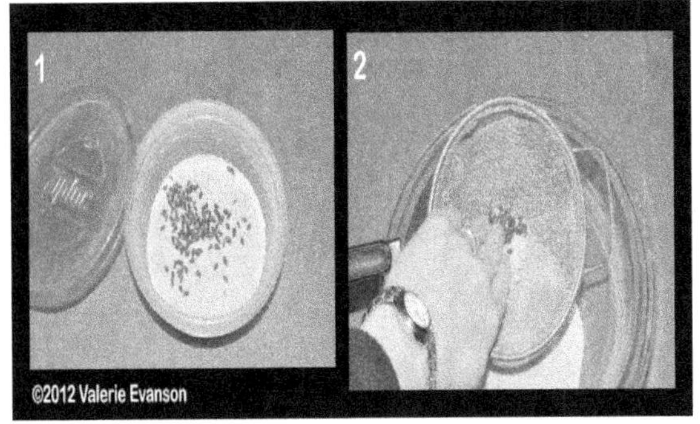

Rubbing seed in sand and separating

Thinning of hard seed coat

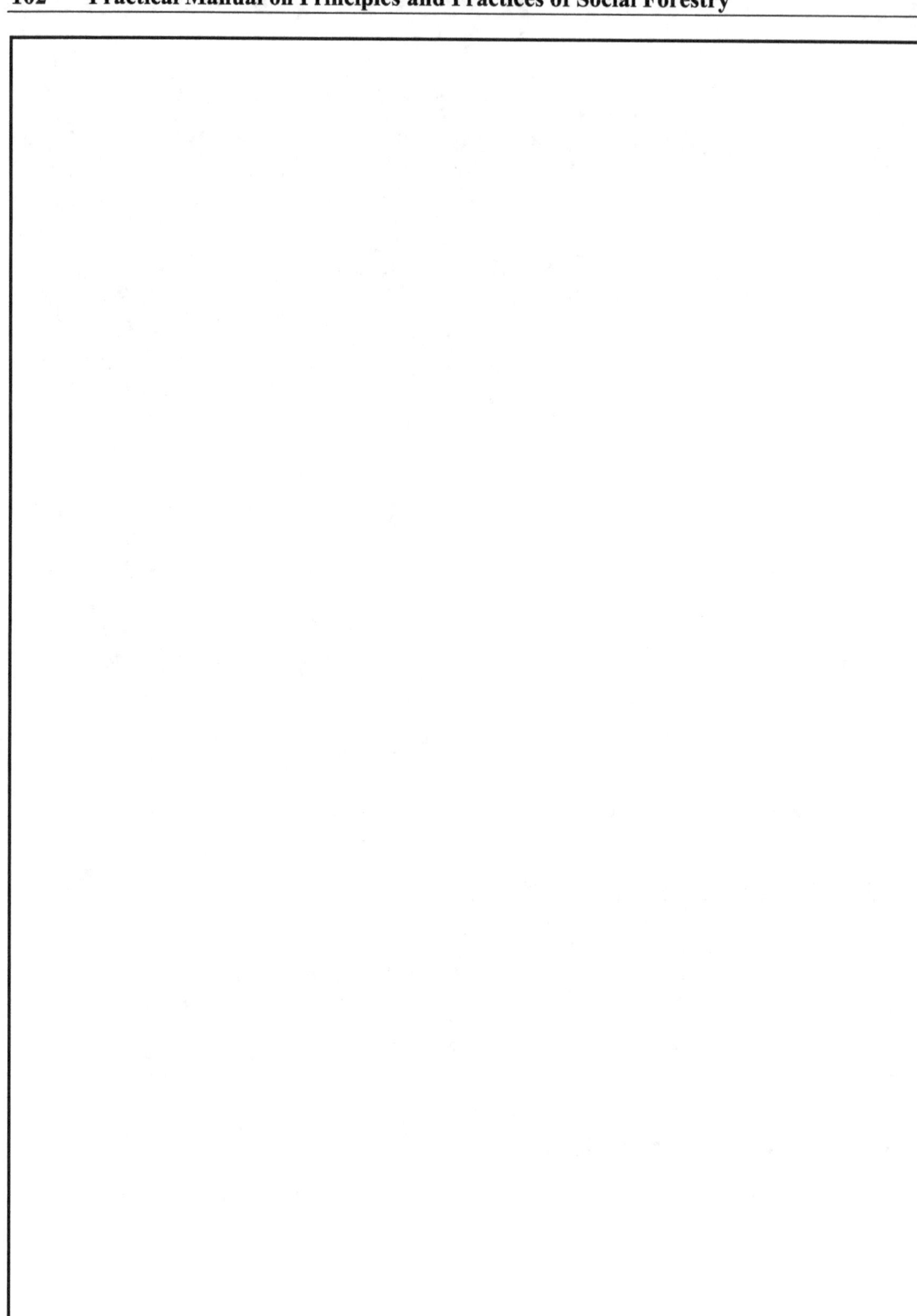

2. Chemical scarification methods

Cold water treatment

Different chemical treatment

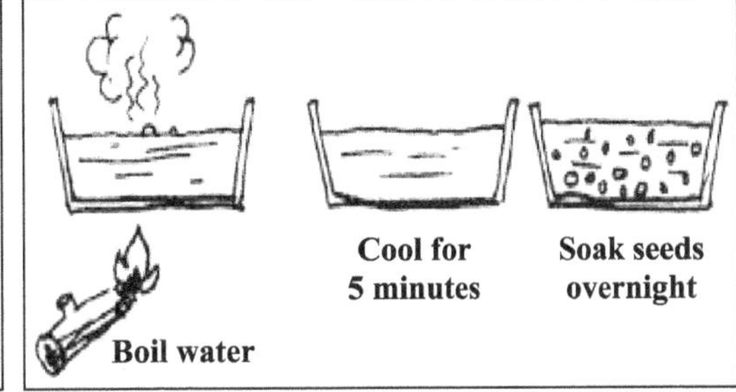

Hot water treatment

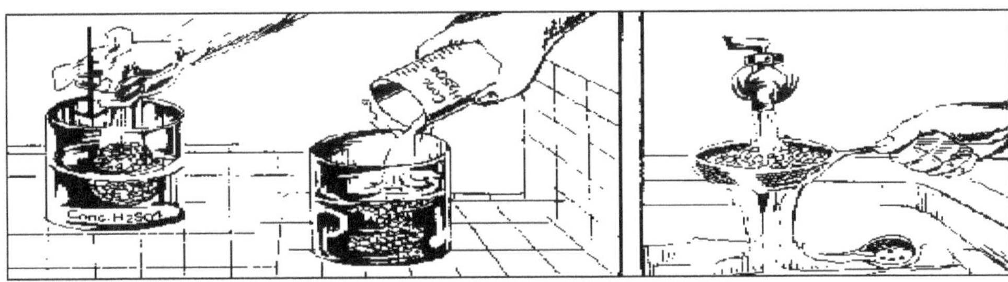

Acid treatment (conc. H_2SO_4)

Soaking in water – alternate wetting and drying

3. Stratification

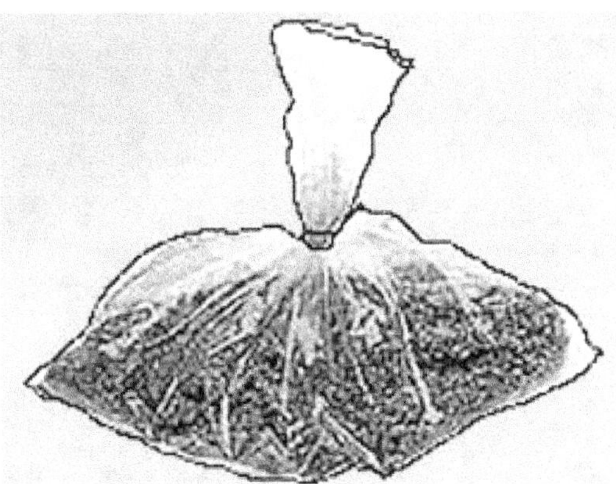

Storing in polybags at temp. 0^0 C – 2^0 C

PREPARATION OF NURSERY BEDS AND SOWING

Objectives:

1. ____

Preparation of nursery beds and sowing

Nursery layout: A well-defined nursery comprises of nursery beds, paths for inspection, site for loading and unloading of material and seedlings and irrigation network etc.

An essential part of the nursery is the nursery bed. Nursery bed is a prepared area for sowing seed (germination bed), transplanting (transplant beds) or root cuttings (rooting beds) or kept fallow (fallow bed) for subsequent sowing / transplanting.

As far as possible the nursery area is to be divided into a number of rectangular blocks (beds) separated by permanent paths. Each block may further be sub-divided into smaller blocks, which in turn may comprise of nursery beds of different types. Irrigation channels may be located in a way that they run along the permanent inspection paths with the main water channels following the boundary of the nursery as shown in figure below.

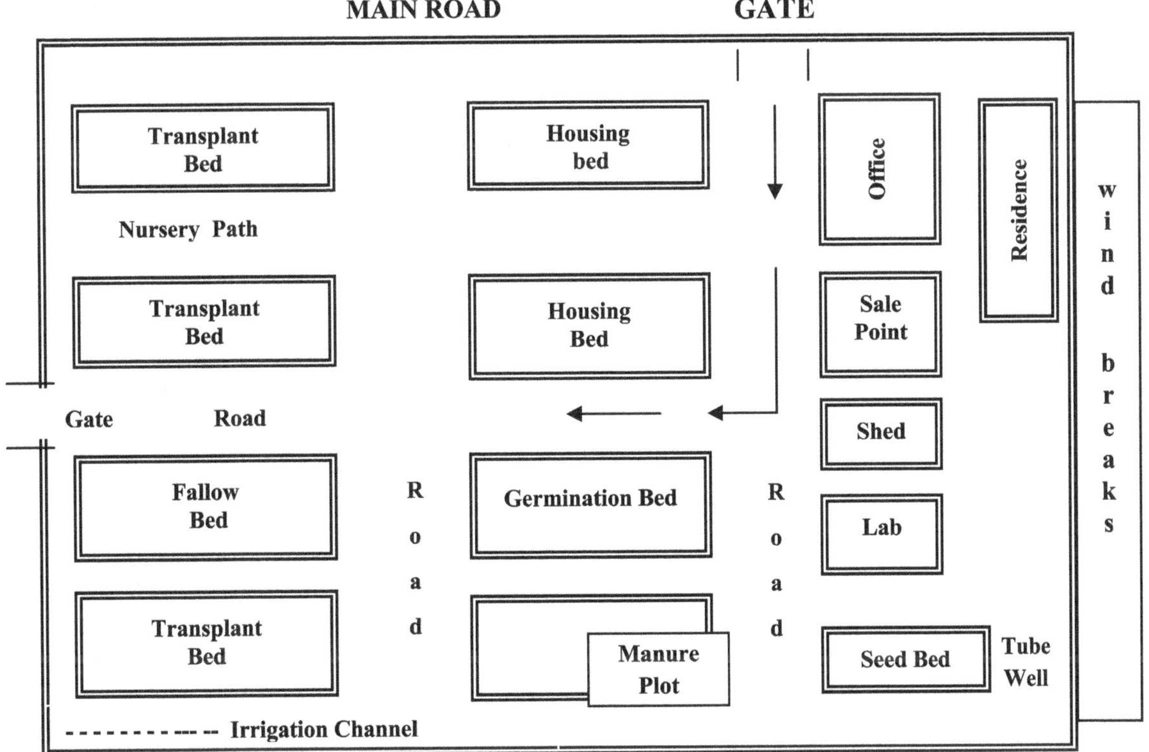

LAYOUT OF A TYPICAL FORESTRY NURSERY

NURSERY TOOLS

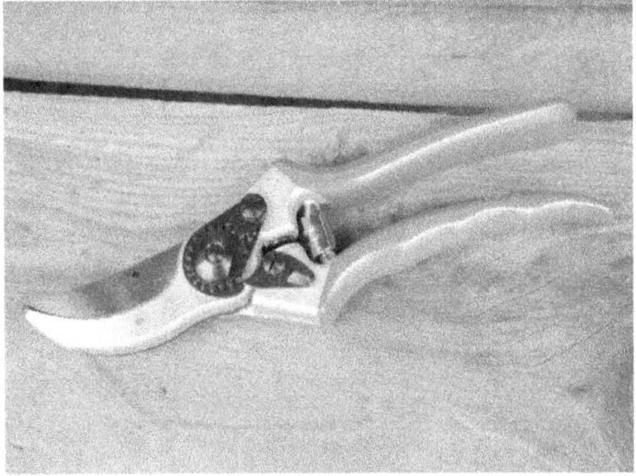

Rose can   Secateur

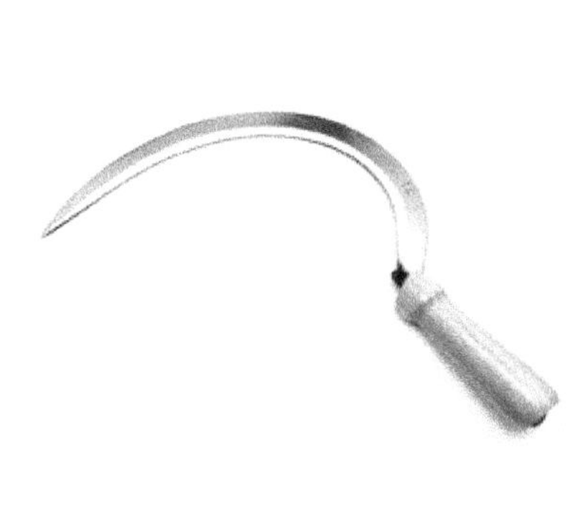

Sickle and Khurpie

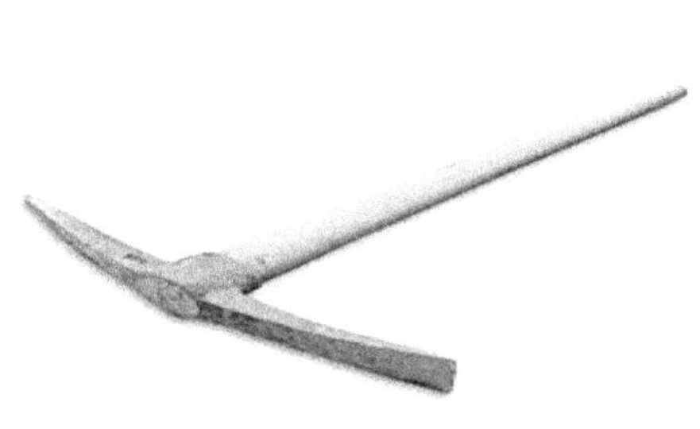

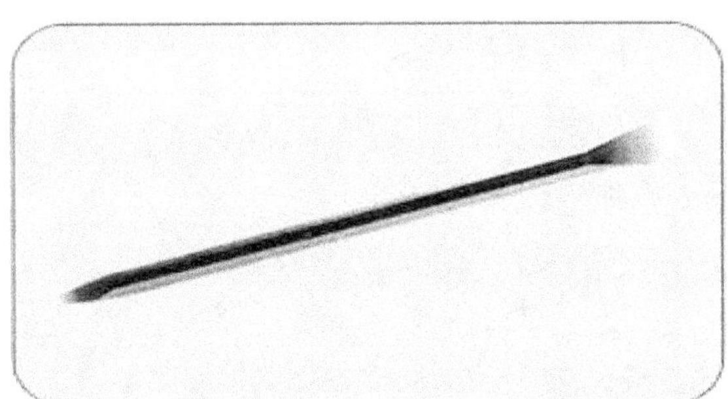

Pickaxe and crowbar

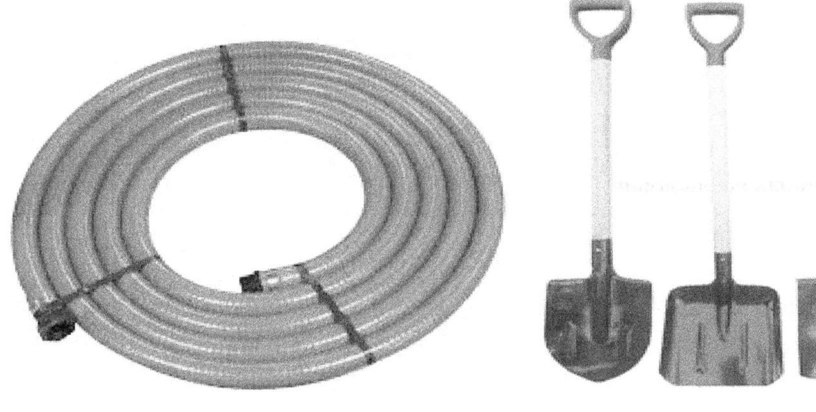

Hose pipe

Shovels and spade

Mini pressure sprayer

Hand sprayer

Nursery layout planning is more difficult in mountain areas where there is a need to build terraces along the slope. Terraces should be wide enough to accommodate both the nursery beds as well as inspection path along it. The beds should be laid out with their length in **East-West direction,** so that they can be shaded against frost or sun.

In case of mechanized nurseries, which feed extensive plantations the path should be wide enough for free movement of heavy equipment such as tractors, bulldozers, trucks, etc.

Preparation of beds

Size of the bed: The size of the nursery bed may vary from (a) species to species (b) locality to locality (c) purpose for which the nursery is being raised.

As a general rule, a bed should have the width which makes it possible for an average man to carryout weeding by sitting on the inspection path without the need to rest his hand or foot in the nursery bed. **Generally 1.25 m wide and 10-15 m long beds are preferred.**

Soil preparation: Soil preparation involves the following steps.

1. The bed is dug upto a depth of about 0.4 m to 1.0 m. All stones, roots etc occurring till this depth are dugout and removed. Digging can be done either manually or mechanically in case of large nurseries.

2. The soil is sieved through a fine wire netting so that all course particles are removed.

3. The fine soil is thoroughly mixed with good farmyard manure. Aldrex may be added if there is a danger of attack of white ants.

4. After sometime this mixture of soil and farmyard manure is again filled back into the pits.

5. The surface of the beds should be either flat or with slant. It may be top dressed with washed river sand, if the soil is heavy, burning of dry grass and shrubby material tends to reduce weed growth.

Types of beds

1. *Seed flats* are used for germinating seeds. They can be in ground made of earthen pots, shallow wooden boxes, plastic trays or bamboo baskets of a portable size i.e., 50 × 30 × 10 cm deep. As the seedlings are to be pricked out in another bed for further growth, the rooting medium in seed flats should allow easy lifting without root damage. The best medium for this purpose is disinfected quartz sand.

2. *Housing beds*: These are used for setting polythene bags. These beds should be 1m in width and are sunk to a minimum depth equal to the length of the container to be kept in the bed. The floor of the housing bed should have a 300-500 microns range polythene sheet to act as a barrier to the seedling roots preventing them to penetrate the earth.

HOUSING BEDS

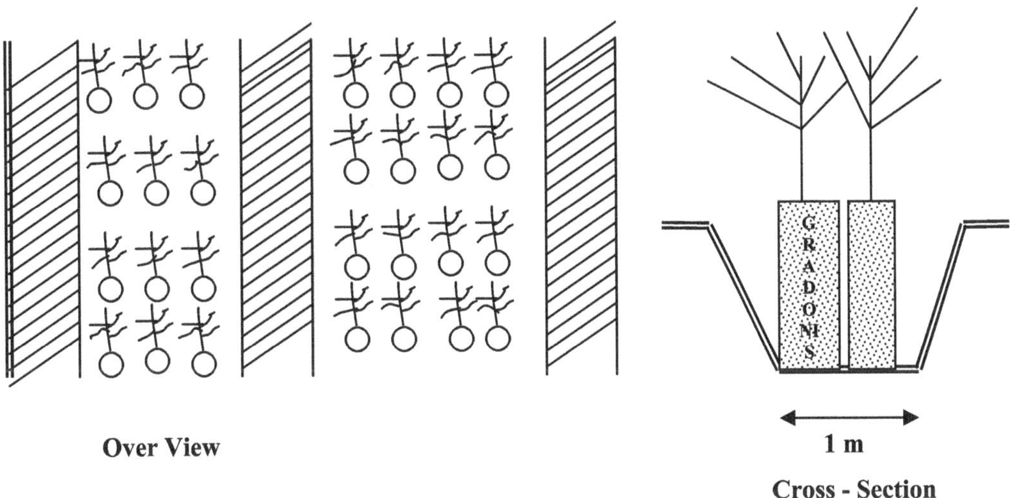

3. ***Raised beds***: In moist areas, which are liable to be water logged, raised beds may have to be prepared. This is done by raising the beds upto 15 cm above the level of the adjoining inspection paths by a line of bricks, stone or even bamboos. This prevents their edges from being eroded away during the rainy season, or the water from seeping into the bed.

RAISED BED

4. ***Sunken beds***: In dry areas there is a need to conserve and collect as much moisture as possible for this purpose, the beds are made a few centimeters below the general level of the inspection path, so that moisture may flow into them. These are known as sunken beds.

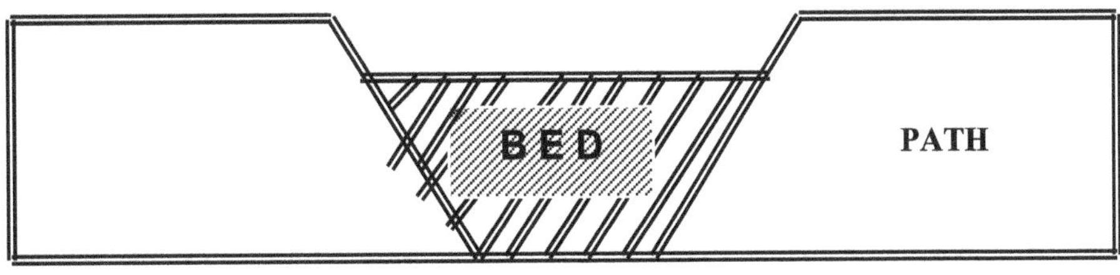

SUNKEN BED

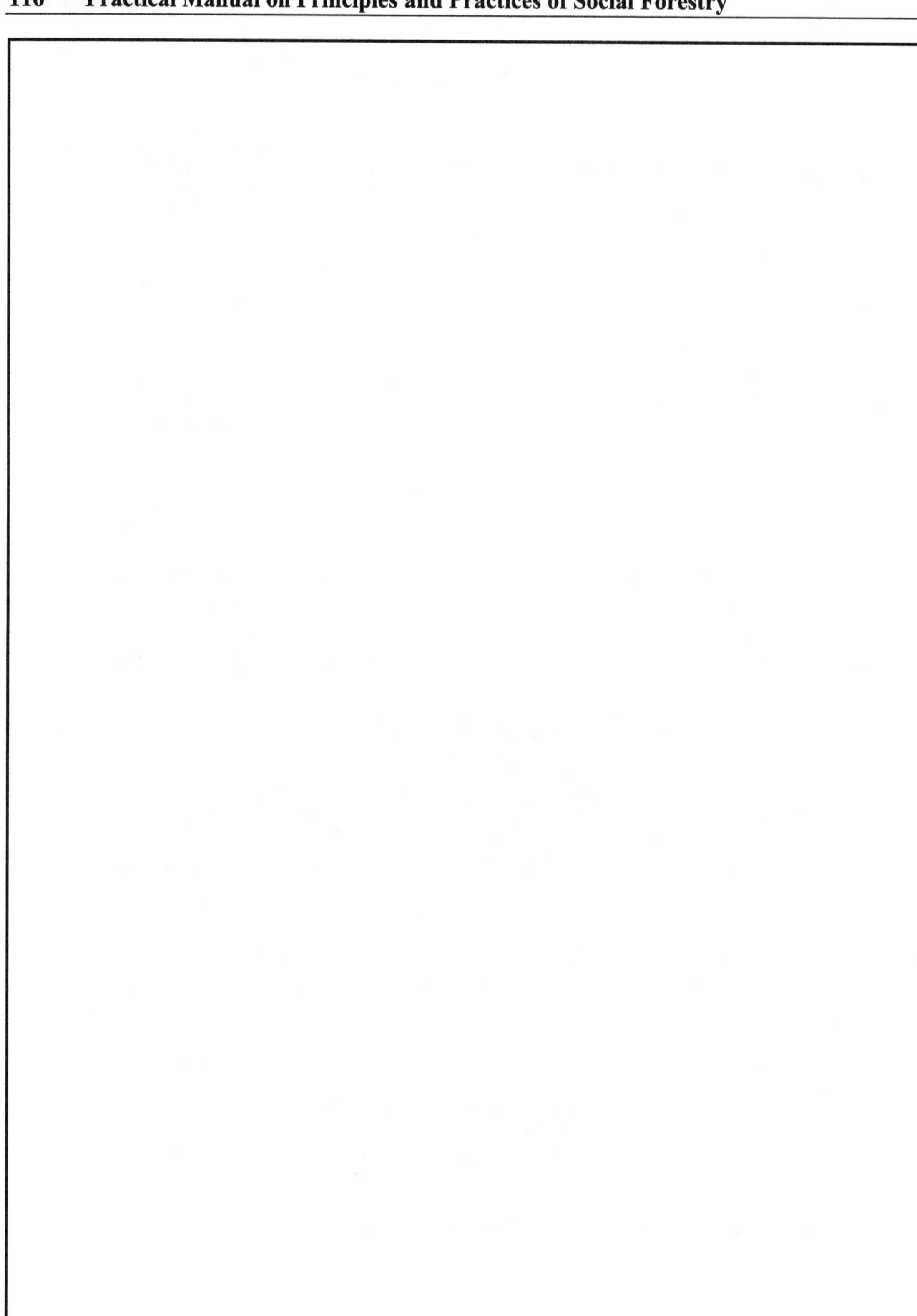

Sowing of seeds: In forest nursery, seed may be sown by any of the following methods.

(a) *Broadcasting*: Seeds are sown by broadcasting it all over the seedling beds. However, this method tends to be wasteful and leads to uneven spacing in the seedling bed.

(b) *Drill method*: This is the most commonly adopted method. Small drills are made in soil in which the seed is placed. This is carried out with the help of a drilling board. It comprises of a board of width of nursery bed of suitable width, with nails or battens being attached to its lower side. In order to make the drills, this board is placed over the nursery bed and pressed. The board is then put over part of the nursery bed and the operation repeated till the whole bed to be sown, has been covered.

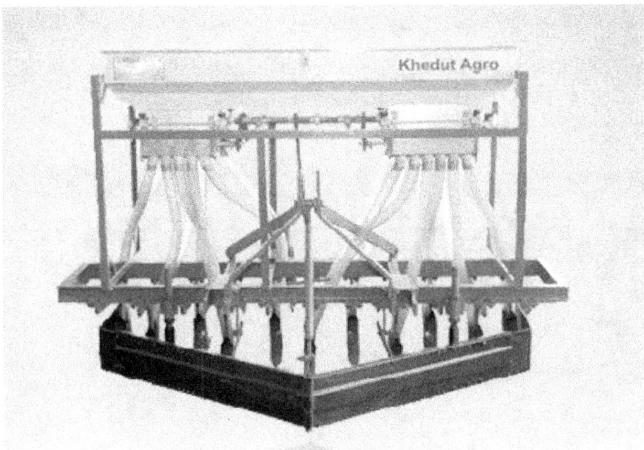

Very often, a hollow drum, having width equal to that of the nursery bed is used for the purpose. The drum is made of tin sheet, to the lower end of which are placed battens for making holes in the soil. The use of such a drum eliminates the need for shifting the drill board from place to place.

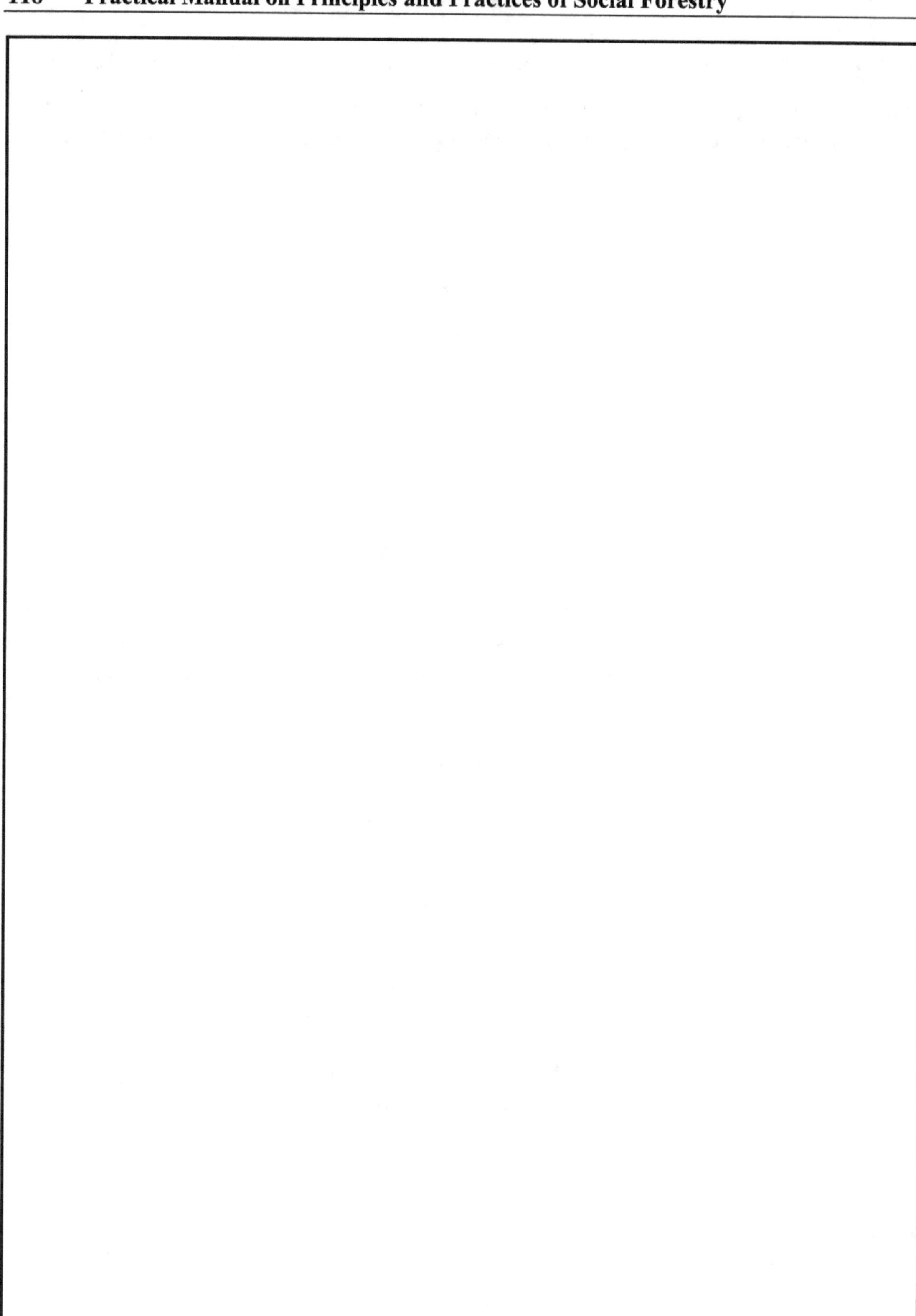

In the absence of a drill board or drum, sowing should be equal to two or three times the diameter of the seed. If the seeds are known to have a comparatively long germination time, the sowing should be deeper, but the soil cover should not be compacted. Seed can be sown more deep in light soils than in heavy soils. Depth of sowing has remarkable effect on seedling emergence and plant growth, as the germinating seed will have to push its tender shoots through the soil above it may die off if the bulk of soils is much more above it. For most of the species a sowing depth of 1.5 to 2.0 cm is normal.

Quantity of seed

The quantity of seed to be sown in each bed depends upon.

1. Weight of the seed
2. Size of the bed
3. Spacing
4. Plant percent

In the above, spacing pertains to the spacing of drills and the spacing of seeds in the drills. The following formula is used to determine the quantity of seed to be sown in each nursery bed. It gives an approximate estimate of the seed quantity in terms of weight.

$$W = \frac{A \times D}{P \times N} \times 100$$

where

W = weight of seed needed in grams

A = Area of nursery bed in sq. mts.

D = No. of plants needed per sq. mts.

P = Plant percent.

N = Number of seeds in each gram.

However, in general practice, between 1.5 – 2 times the quantity of seed should be used in case of drill sowing and 5-6 times in case of broad cast sowings.

Time of sowing

The time or season of sowing depends upon the following factors.

1. Time or season of ripening of the seed.
2. Rate of growth of the species, which is being raised.
3. Size of the plant to be transplanted.
4. Climatic conditions of the area or locality.

Depth of sowing and seedling density in nurseries

S. No.	Species	Optimum depth of sowing (cm)	Density [spacing in nursery (cm)]	Germination percent	Optimum spacing in transplant beds (cm)
1.	*Acacia catechu*	1.5	20 x 20	60-80	5 × 20
2.	*Acacia nilotica*	1.5	15 x 15	50	5 × 20
3.	*Adina cordifolia*	0.5	15 x 10	30-40	20 × 10
4.	*Ailanthus excelsa*	0.5	20 x 10	70-90	20 × 10
5.	*Albizzia lebbek*	1-1.5	15 x 12	60-94	25 × 10
6.	*Azadirachta indica*	1.0	20 x 5	40	15 × 15
7.	*Butea monosperma*	(L.C)	8 x 15	73-90	15 × 15
8.	*Dalbergia sissoo*	1.5	25 x 2	90-100	10 × 10
9.	*Grewia optiva*	2.0	10 x 2	50	15 × 10
10.	*Holoptelia integrifolia*	(L.C)	10 x 10	60	20 × 10
11.	*Leucaena leucocephala*	1.2-1.5	30 x 5	80	10 × 10
12.	*Madhuca longifolia*	1.2-2.5	15 x 3	85	30 × 15
13.	*Melia azaderach*	2.0	15 x 10	70-80	15 × 15
14.	*Pongamia pinnata*	2.0	15 x 2	80	30 × 15
15.	*Gmelina arborea*	1.0-1.5	15 x 10	43-85	15 × 20
16.	*Bombax ceiba*	(L.C)	15 x 5	14-75	25 × 20
17.	*Eucalyptus tereticornis*	(L.C)	5 x 2	90	5 × 10

(L.C) = Light covering)

Covering the seed and nursery beds

As soon as the seed has been sown, it should be covered by the soil, in a manner indicated above. In case where there is a danger of the seed being eaten by birds, rodents, pets etc., the following protective measures may be adopted.

(a) *Application of insecticide*: Insecticides or insect repellents may be added to the soil. For eg. Aldrin, Aldrex, Methyl parathion, Folidol dust etc., may be mixed with the soil while preparing the nursery beds. This acts as a sort of repellent and will protect the seed from attack by insects.

(b) *Covering the seed bed*: The seed bed may also be covered by thorns and dry bushes, to protect the underlying seed from being devoured by birds and rodents. In advanced modern nurseries, caging by wire netting is practiced.

(c) *Application of repellent on seed*: Very often insect repellent may be mixed with the seed eg: red lead, kerosene oil and camphor. Liquid repellents may be thoroughly rubbed on to the seed.

TRANSPLANTING, RAISING OF BARE-ROOTED AND CONTAINERIZED PLANTS

Objectives:

1. ____

2. ____

3. ____

Transplanting: Seedlings of certain species such as chirpine are kept in the nursery for a period less than a year. These can be transferred from the seedbed to the field. This is known as planting out. However, when the seedlings have to be kept in the nursery for more than a year, it must be transferred to beds other than the seedling beds. This is known as pricking out or lining out or transplanting out. If seedlings are allowed to remain in the seedling bed for a long period, they are likely to develop a long taproot, thus making it difficult to transplant them in the field.

Pricking out forms a very important operation in case of species that have to be kept in the nursery for more than a year. In case of species such as fir and spruce, which have to be raised in the nursery for 3 or 4 years or even more, pricking out is carried out each year.

Advantages:
1. Seedlings can remain in the nursery till they are of a size big enough for planting in the field.
2. They are able to get enough space for development. Seedlings are prevented from developing a long tap root, making it difficult to plant them in the field. A fibrous root system develops if pricking out is carried at early intervals.
3. It enables easy grading of the plants, particularly in cases of species that are to be raised in the nursery for 3 years or more.

After removing seedlings from seeding beds they may be transferred to (a) transplant beds at a wider spacing or (b) planting bricks, polythene bags or other containers.

Procedure: Pricking out involves the following steps
1. Small trench is dug along one end of the seedling bed.
2. The first row of seedlings are then pushed into the trench along with amount of soil, by driving a shovel on the side of the plants away from the trench in the ground, deep enough and pushing it towards the trench. Thus the plants are dislodged from the nursery bed, without any damage to the root system.
3. The plants are then transported to the transplant beds.
4. Transplanting is done with the help of a transplant board, which has notches at the desired spacing, so as to enable the plants to be paced into the soil.
5. Once transplanting has been done, watering is carried out by means of watering can. Before removing the transplanting boards, the soil is properly pressed.

Mechanised transplanting

In developed western countries most of the nurseries have transplanting machines that may transplant 5-7 rows simultaneously at even spacing. 3-5 persons sitting on the carriage with bundles of uprooted seedlings will be feeding these seedlings in the grooves of rotating wheels and as the machine moves forward the seedlings get transplanted.

Wrenching and undercutting

Wrenching is cutting of side roots by working small machines on both sides of the seedling rows. This will promote development of more fibrous roots and help in better transplanting. The machines employed for this purpose are called wrenchers. Under cutting is cutting the taproot at certain depth. This can be done by an implement that has a sharp blade and is worked at the required depth that cuts the taproot of the seedlings in the bed. Both wrenching and under cutting should be done 3-4 weeks before the actual transplanting to be taken up.

Production of bare-rooted plants

All the operations up to transplanting are common for production of both bare-rooted and containerized plants. For production of bare-rooted plants the seedlings are transplanted in the beds at a wider spacing to allow sufficient space for growing upto desired height and girth. If much taller plants are required they are again transplanted into a secondary bed by increasing spacing after one year for promoting growth. The following operations are done regularly.

(a) *Weeding*: This is an important operation to be done regularly both in seedling beds and transplant beds. They offer competition to the struggling seedlings for light and food, hence have to be removed from time to time. In the initial stages, the weeds are small and can easily be just pulled out without disturbing the soil. For grown up weeds iron implements with sharp edge may be used.

The following precautions need to be adopted during weeding operations.

1. Care should be taken to ensure that the labourers are fully able to recognize the seedlings of the desired species and does not pull it out or harm it in any way, during weeding operations.

2. The roots of the desired seedlings are in no way disturbed. This can be brought about by holding the seedlings firmly in place between two fingers, while the weeds are removed by the other hand.

3. The labourers should not enter or sit on the nursery bed. Weeding should be done by standing or sitting on the adjoining inspection path.

4. Weeding should not be done when soil is wet.

5. In case the seedlings of desired species have become too crowded they may be lightly thinned during weeding.

6. A light working of the upper soil surface of the nursery bed may be carried out. In the transplant beds weeding can be done mechanically to reduce cost by working implements, weeds, push hoes etc.

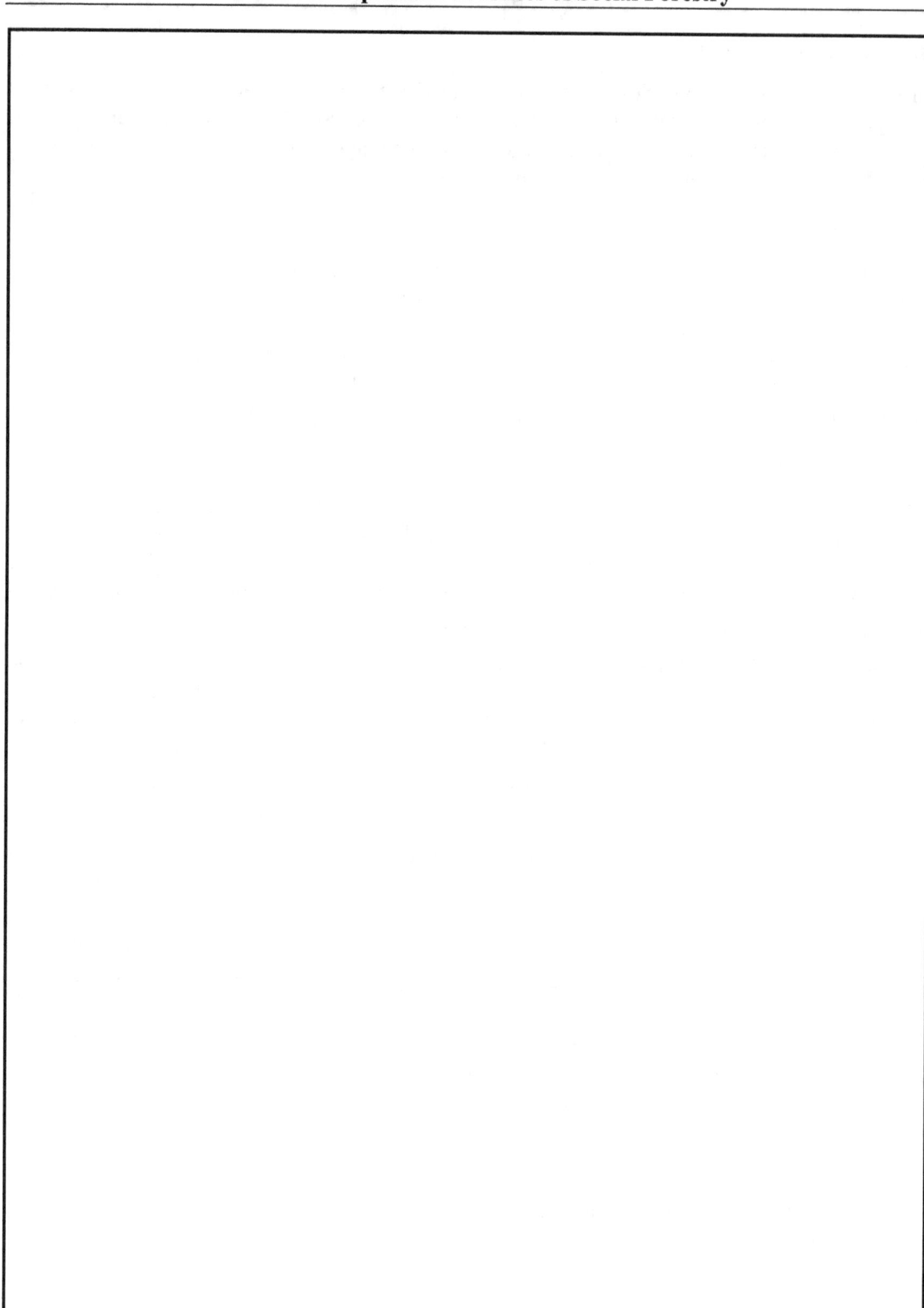

(b) Shading

1. *Protection against the sun*: Young seedlings of certain shade bearing species may need shade against the excessive heat of the sun, during the pre-monsoon period.

2. *Protection against frost*: In frost prone areas the shade is to be removed each day so that the soil temperature rises sufficiently and adverse effect of frost is reduced.

(c) *Irrigation*: Nurseries may be irrigated by the following methods.

1. *Can irrigation*: Watering cans bearing finely perforated roses may be used for watering which gives the effect of natural rainfall. It is suitable for small nurseries.

2. *Automizer irrigation*: An automizer is a pump by which moisture is applied to the seedbeds in the form of water vapour. Excess moisture is likely to   damage seedlings of species such as eucalyptus. The automizer enables water to settle down upon the seedbed as a spray, thereby reducing the danger of soil erosion. Seedlings may be irrigated by automizers till they attain a height of 10-15 cms, after which the usual methods of irrigation may be adopted.

3. *Percolation*: Under this method water is made to percolate to the center of the nursery bed from the sides. It ensures that the nursery bed does not form a crust.

4. *Sprinkler irrigation*: In this method, water comes out in the form of small jets from pipes that are laid out all over the nursery bed. Even though an extra investment is required, this method is good for economy in water and labour.

5. *Flood irrigation*: This method involves flooding of the beds with water. It is normally adopted in dry arid areas with sunken beds. The nursery bed is flooded with water to a depth upto 5 cms. The water is allowed to remain in the seedbed for quite some time depending upon the local conditions. This method is not suitable in areas having water shortage or when seeds and seedlings are slow growing.

6. *Hardening off*: This is a process in which seedlings are to be planted within a short period of time are hardened off or conditioned for field conditions, by slowly reducing water, shade, shelter etc.

In nurseries with shaded beds, the young seedlings become used to conditions of shade. Hence, before transplanting they must be conditioned to withstand open conditions in the forest. This is done by slowly reducing the shade. In the absence of hardening off, plants are likely to be killed after planting.

Production of container plants

The production of containerized plant involves all the steps needed in the production of bare noted plants. Instead of planting the seedlings with naked roots in the forest, they are first established in containers of different types before planting in forest to reduce casualities. Alternatively the transplanting damage is altogether eliminated if seeds are directly sown in the containers. Failure of germination of seed will however, involves increase in expenditure. To overcome this sowing of pre-sprouted seeds is recommended for chirpine.

The greater disadvantage of using containers is that there is always risk of root coiling when the taproot is overgrown. Plants with such coiled root system do not survive long under field conditions.

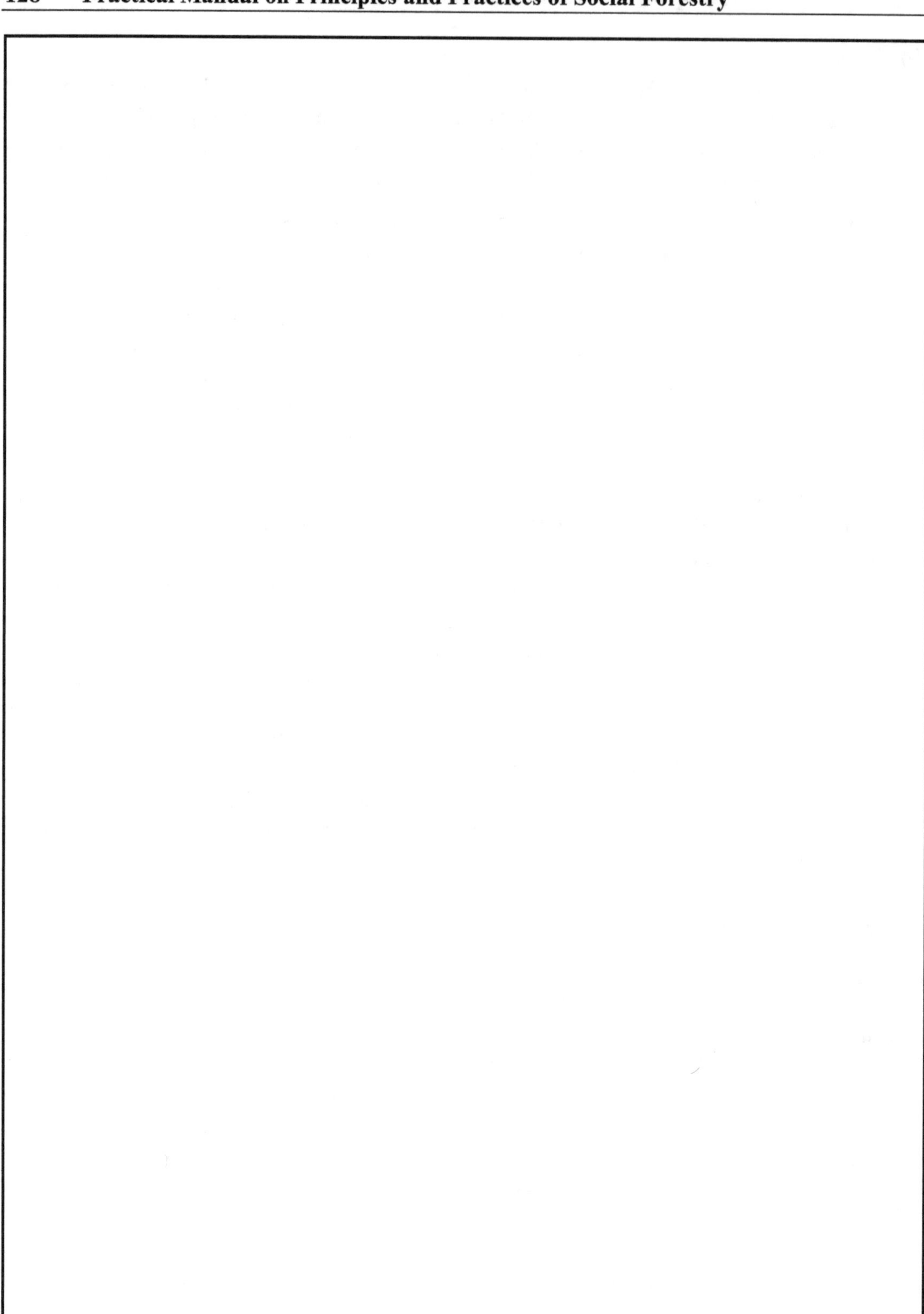

To avoid this problem it is advisable to provide a circular hole of 1.5 cm diameter at the bottom to allow the tap root to pass through the hole into the underneath soil. At that time cut with a sharp implement. This will avoid development of deformed root system after planting in the field and will also encourage the development of long tap root that is necessary in successful establishment of plants in places experiencing long dry spells.

Types of container used

1. ***Transplanting bricks***: Transplanting or planting bricks are used to facilitate the transport and translating the seedlings in the field. They comprise of unbaked or kacha bricks made up of clay, farm yard manure and sand in a ratio (1:1:1). Water is added to make the mixture semi viscous, after which it is poured into a specially designed mould. Planting bricks are narrow at the top and gradually widen towards their base. The commonly adopted dimensions are;

Height 25-30 cm, top 10 × 10 cms and base 15 × 15 cms.

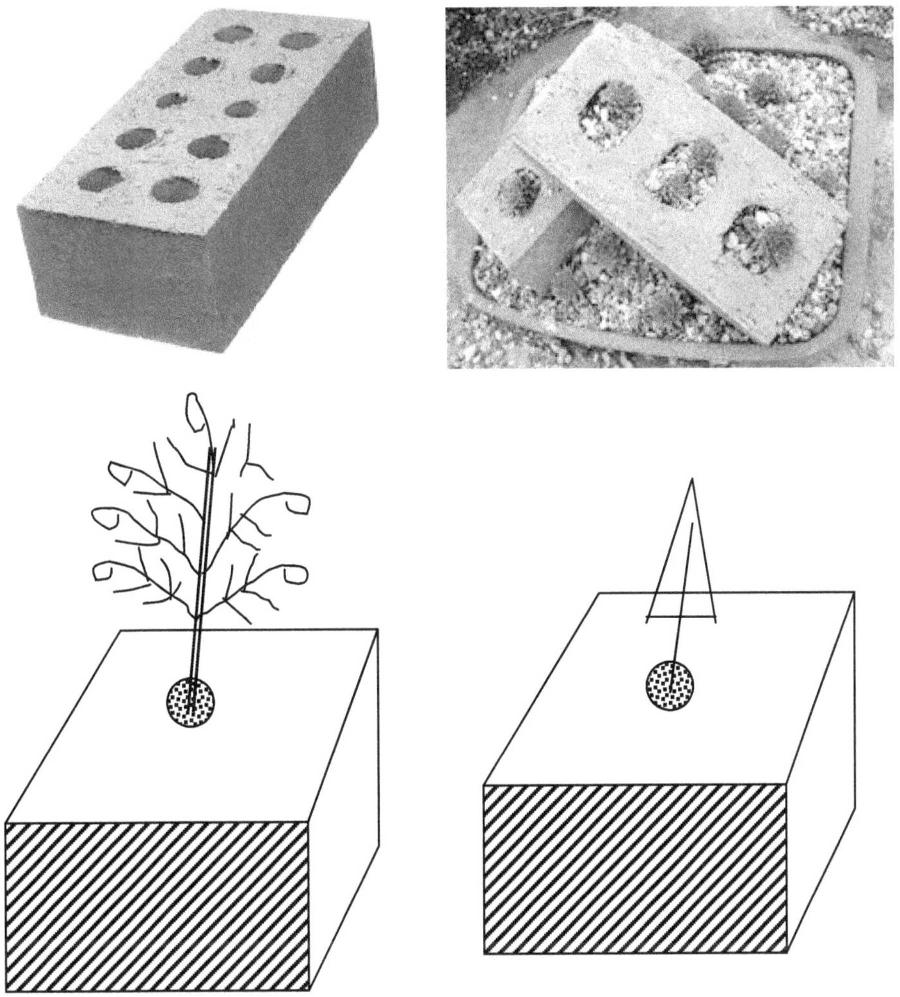

PLANTING BRICKS

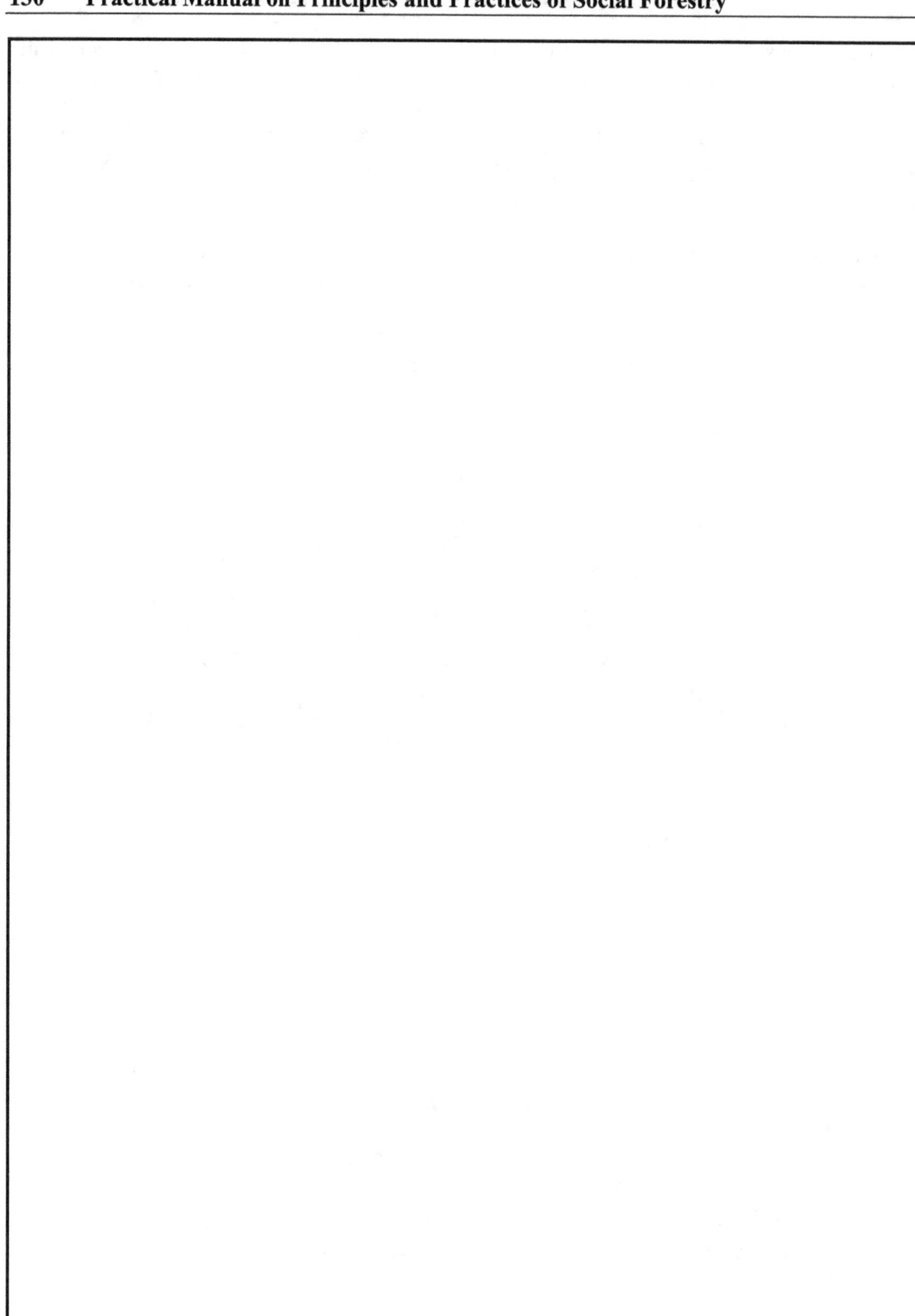

A small hole, 2 cms in diameter and upto 8 cms deep is left at the top center of the brick. This is meant for sowing seed or transplanting the seedling. Once the planting brick has dried out, this hole is filled with nursery soil and farm yard manure 1:1 proportion. Very often, the entire planting brick bearing the plants is placed into the nursery bed and its edges covered with soil. The base of each brick thus placed may be over 2 cms apart. Nursery beds containing the planting bricks may be treated as usual.

2. ***Dona containers*:** This is local container made up of leaves of ficus, bauhinia or *Butea monosperma,* which is cup shaped. They are filled with soil and FYM and plants placed into them.

3. ***Baskets*:** Baskets made of bamboo strips or tamarix may be used as containers. Soil is placed into them. While transplanting they may be buried into the soil together with the plant.

4. ***Moss cylinders and fiber cubes*:** Moss may be made into rough cylinders and plants placed into it. These are used for eucalyptus in ooty hills. Cubes made up of fibers may also used for the purpose.

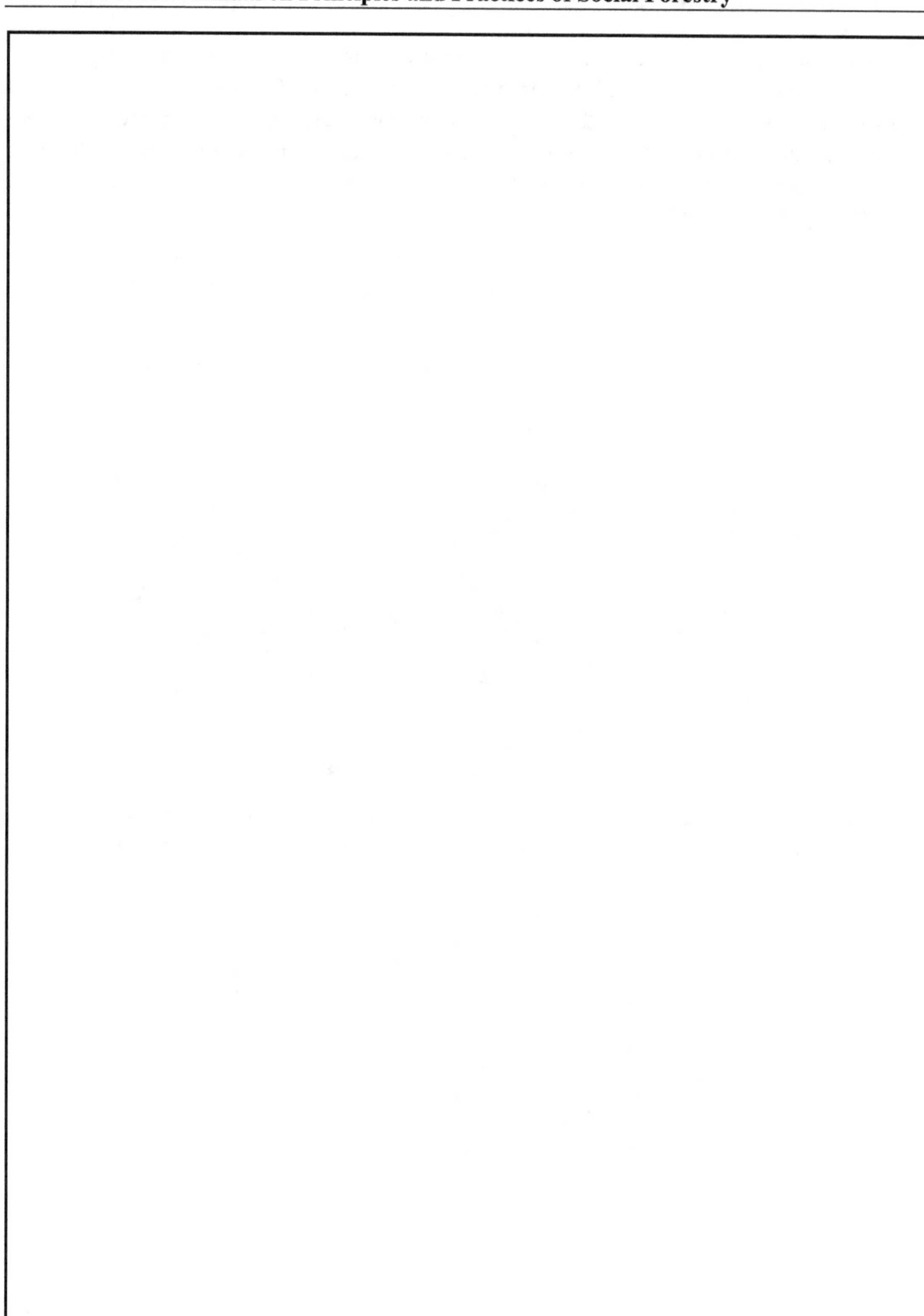

5. *Tubes*: Suitable containers in the form of tubes made of earthern tiles, bamboos, tin and betula bark.

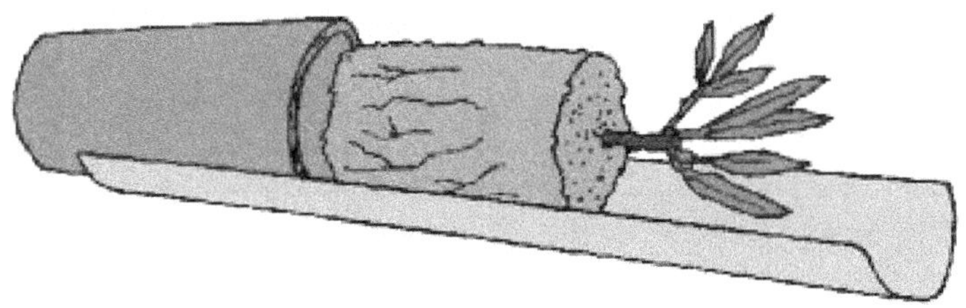

6. *Earthern pots*: With or without bottom may also be used as containers.

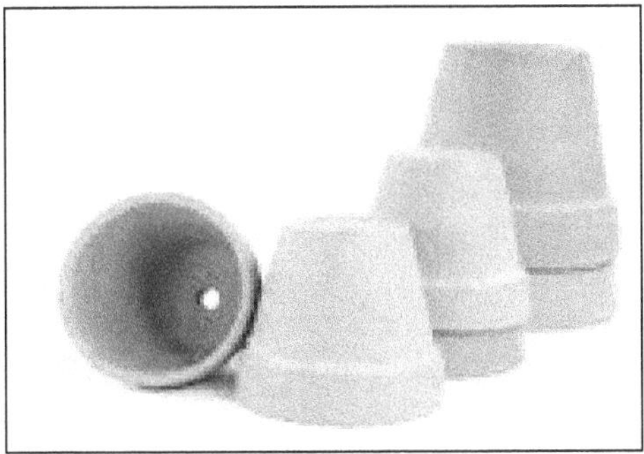

7. *Polythene bags*: These are most commonly used containers in the country. They comprise of a cylindrical tube of polythene.

Polythene Bag Planting

Nursery diseases

Damping off is the main disease affecting the seedlings in a nursery. Seedlings are attacked at their base near ground level by fungi belonging to Phytophthora, Pythium, Rhizoctonia and Fusarium which are usually found in the soil. Due to fungus attack, rotting starts at the base of the seedlings and they may suddenly fall on the ground.

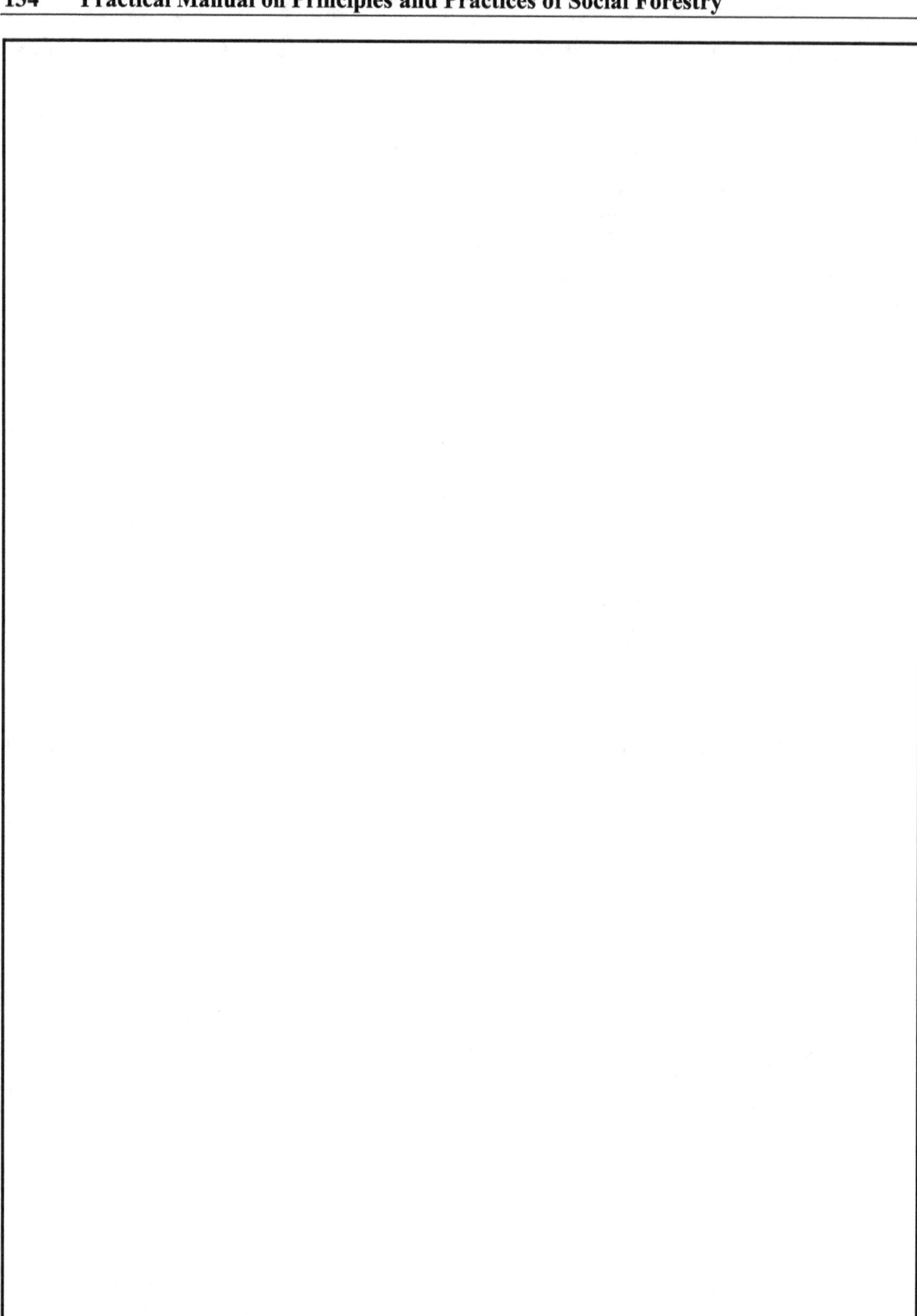

134   **Practical Manual on Principles and Practices of Social Forestry**

Measures to control nursery diseases

1. *Soil conditions*: Various diseases may be kept at a low level by growing in light texture, well drained acidic soils. In case the soil is nearly neutral or alkaline, it may be acidified by (a) concentrated H_2SO_4 (100-400ml in 45 litres of water) (b) sulphur (250-300g) (c) Aluminium sulphate (powdered, iron free 17-18% 340-680 g) (d) Ammonium sulphate fertilizers. The exact dosage of the chemical to be applied will depend on the buffering capacity of the soil.

2. *Steaming*: This is a process in which steam is passed through the nursery soil. As a result of the higher temperature at which steam is passed, the soil pathogens are killed. There are two methods of steaming the soil (a) soil may be under running steam in an autoclave or closed container for an hour or so. (b) In seed beds, pipes with perforations may be laid into the beds which are covered with gunny bags. Steam is let into those pipes for an hour or so in each bed.

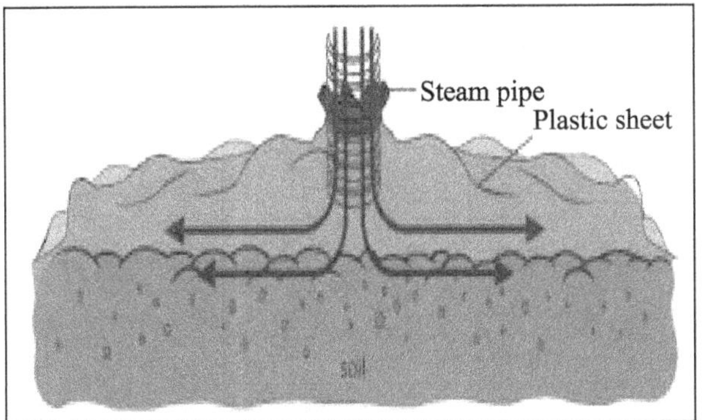

3. *Chemicals*: A number of chemicals are used for treatment of the nursery soil against diseases. Since they may be phytotoxic they have to be applied 10 -15 days in advance of sowing. The chemicals commonly used are formalin, methyl bromide, blitox, captan, zineb etc.

4. *Fungicidal seed treatment*: This involves the use of chemicals like, thyride, blitox etc. Fungicidal seed treatment brings about effective control of damping off and also seed borne diseases in case of a member of broad-leaved species.

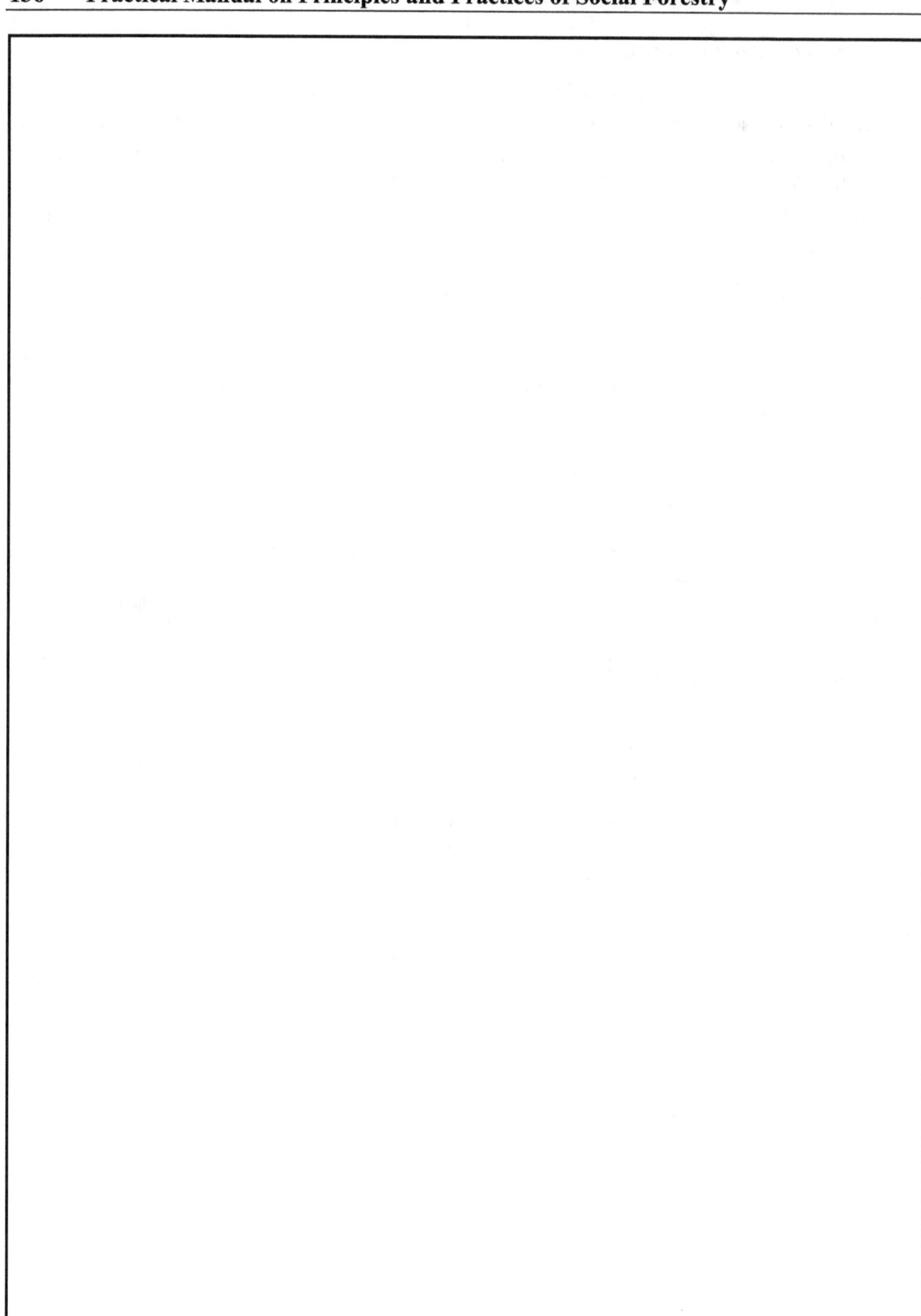

FIELD PLANTING TECHNIQUES AND AFTERCARE (TENDING)

Objectives:

1. ____

2. ____

3. ____

Soil preparation: Soil preparation pertains to working the soil in the plantation area at places where sowing or planting is to be done. The soil may be dug either by manual labour or by mechanized means.

1. ***Pits***: Pits are dug for purpose of planting saplings. They are of the following main types.

   (a) ***Ordinary Pits***: This is the normal type of pit 30 × 30 cms at the base, 45 × 45 cm at top and about 30 cm deep. It is provided with a high crest and a ring for storage of moisture.

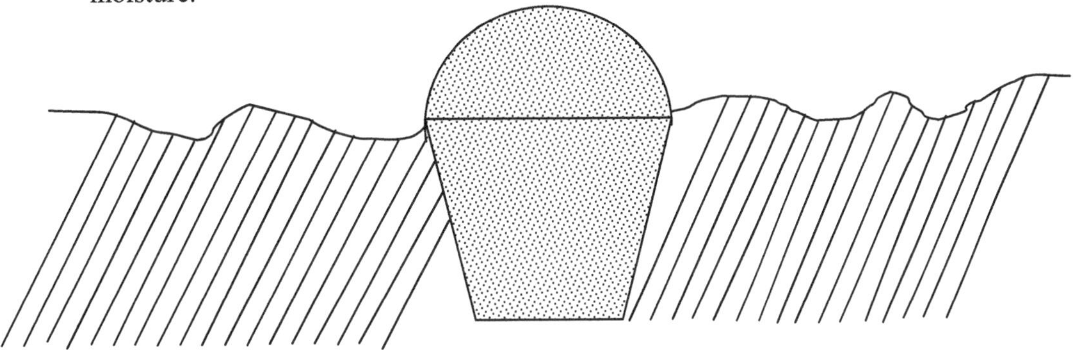

ORDINARY PIT

   (b) ***Saucer Pits***: This is a pit dug up in the center of a saucer shaped structure of one meter radius, prepared by giving the soil a gentle slope inwards from the outer periphery.

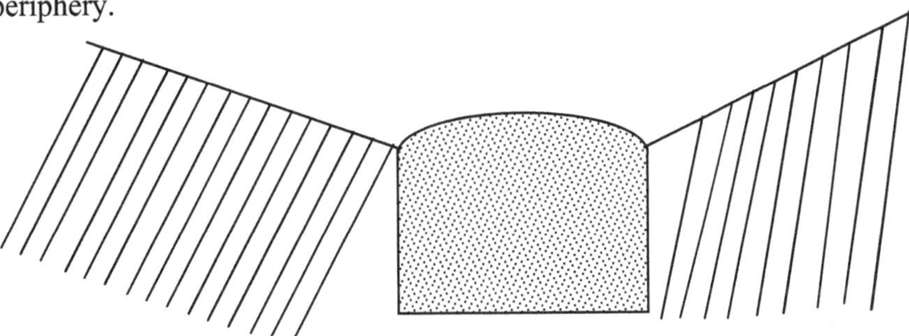

SAUCER PIT

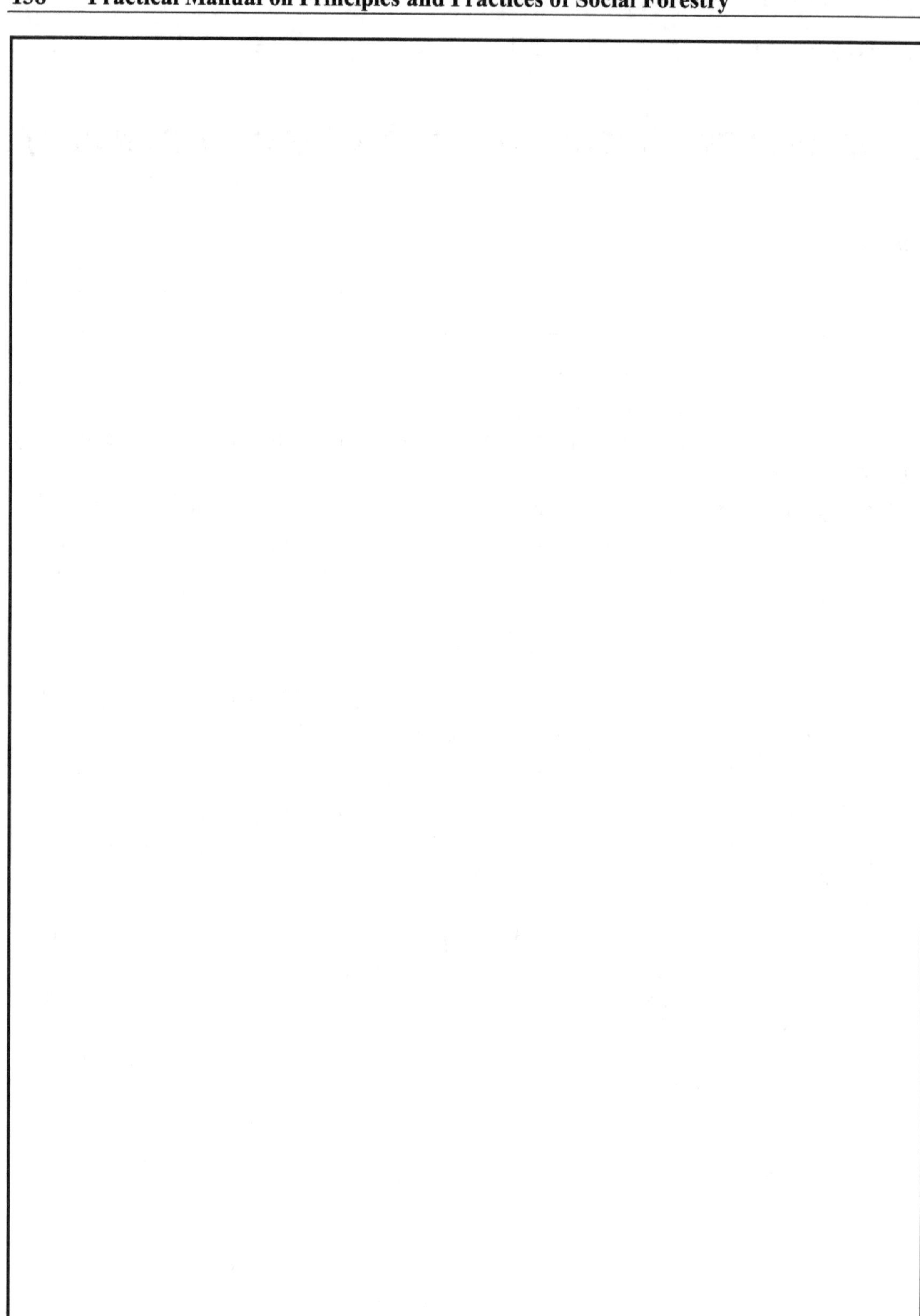

(c) *Crescent pit*: This is a modification of the saucer pit. In this type, the pit is crescent shaped.

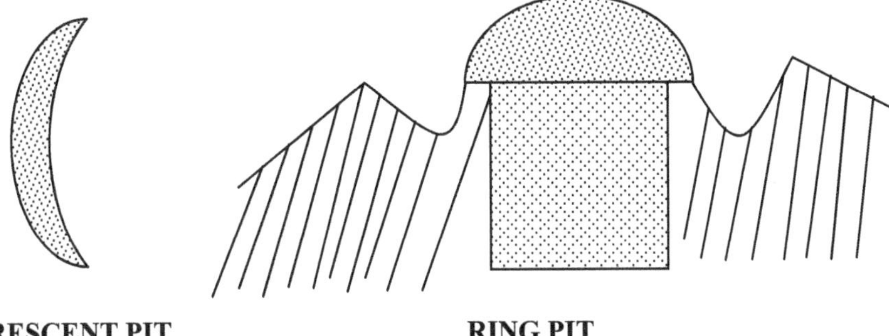

CRESCENT PIT   **RING PIT**

(d) *Ring pit*: This is a pit which has a circular trench about 20-30 cm away from it. The main purpose of this circular trench is to collect rainwater for supply to sapling.

2. *Trenches*: Trenches are dug for purpose of planting. Main types of trenches are

(a) *Ridge ditch*: It is a partly filled trench with a ridge along it. These are suitable for sloping grounds in which the ridge is made on the lower side of the trench.

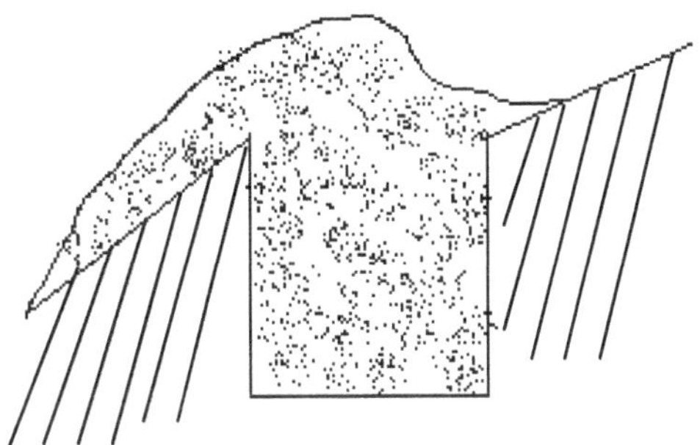

RIDGE DITCH

(b) *Shelfed trench*: This is a type of trench with a shelf at one end and suitable for clayey soils forming an impermeable crust on wetting and drying.

(c) *Double trench*: It comprises of two trenches, one is filled with soil and the other remains empty. The latter is used for storing water and hence is made on the upper side of the slope, whereas the filled trench may be used for the purpose of sowing or planting.

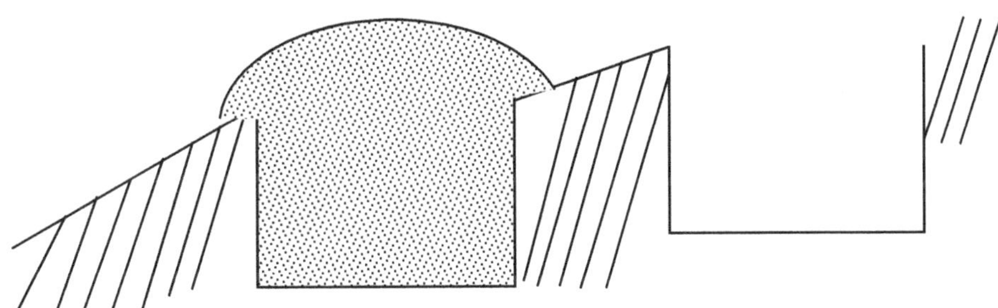

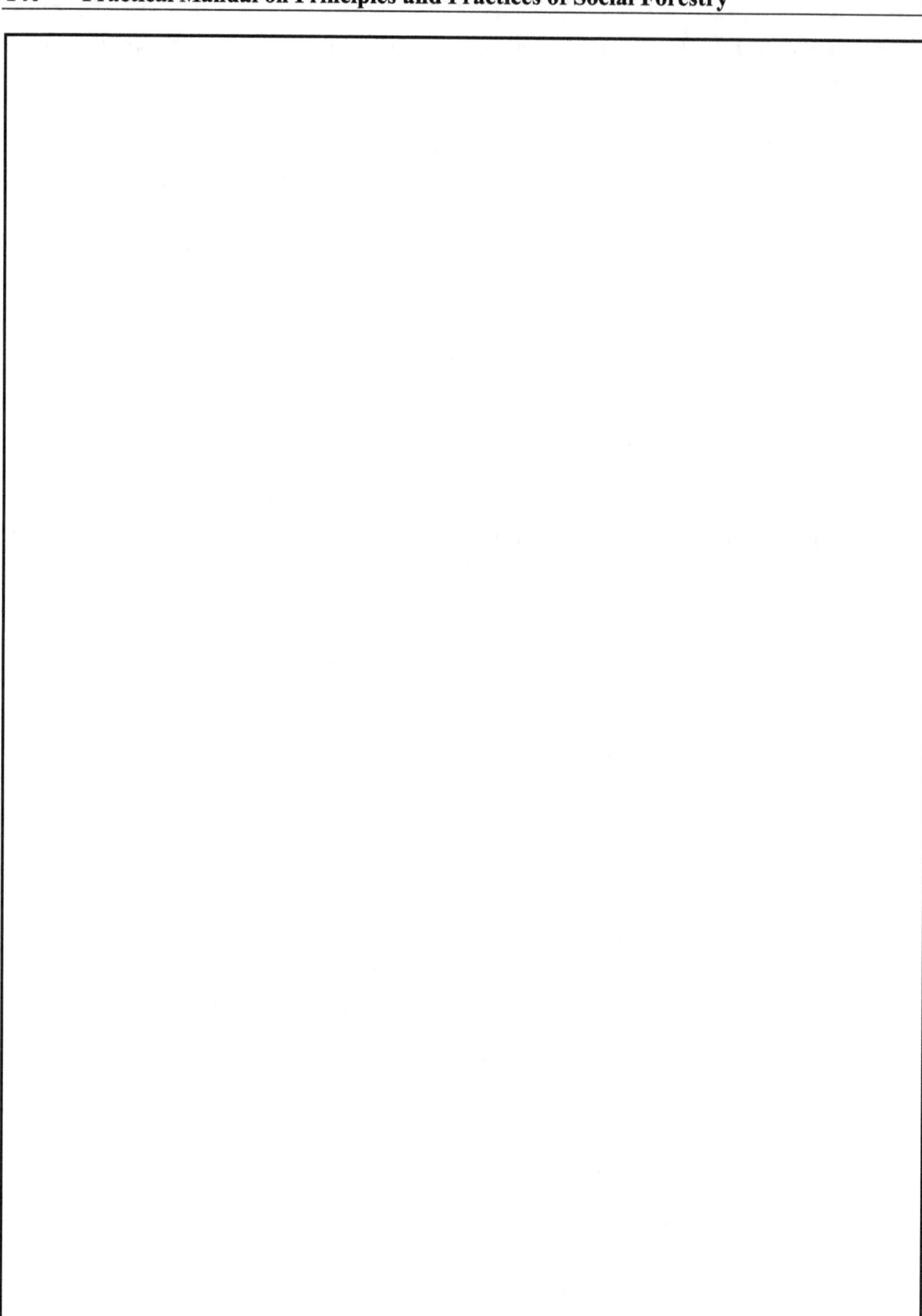

(d) *Inclined trench mound*: In this type an inclined mound is made in the middle of the trench.

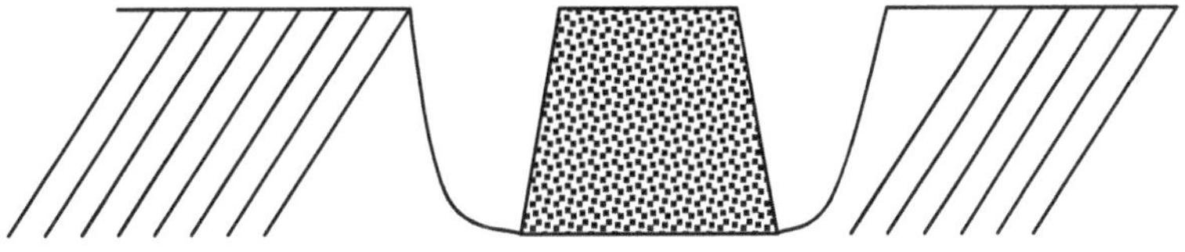

INCLINED TRENCH MOUND

(e) *Wurli*: It is a kind of shallow trench ridge and is suitable only for gentle slopes with loamy or granular soil in areas with rather high and well distributed rainfall. The worked out soil is heaped so that a shallow trench is available on the upper side.

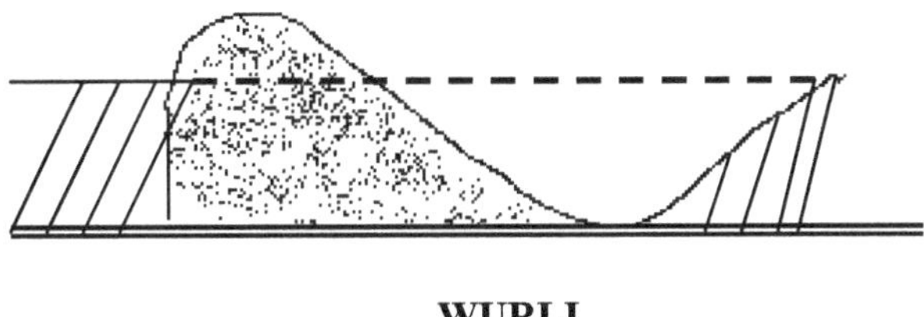

WURLI

3. *Mini catchments*

(a) *For level lands*: For *in situ* water harvesting on level lands, depressions resembling the shape of saucer of 15 cm depth at the center and minimum diameter of one meter are made. Tree is planted in the center of saucer.

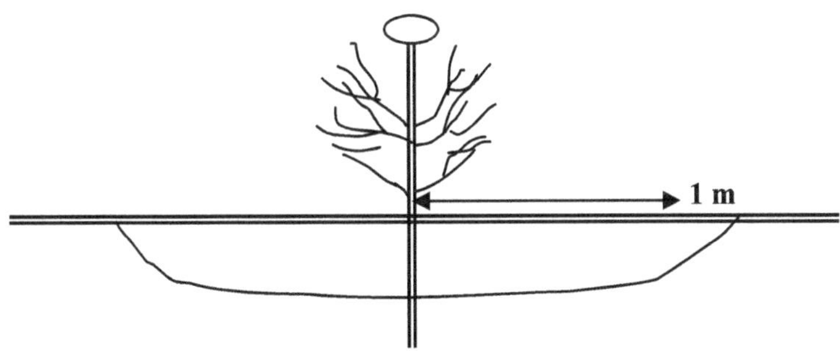

SAUCER SHAPED MINI - CATCHMENT

(b) *For mild sloppy land (2 - 4 % slope)*: Depression resembling the shape of 'V' of 15 cm depth with a minimum arm length of one meter are made and the tree is planted near the apex of 'V'. Half moon shaped depressions are also practiced.

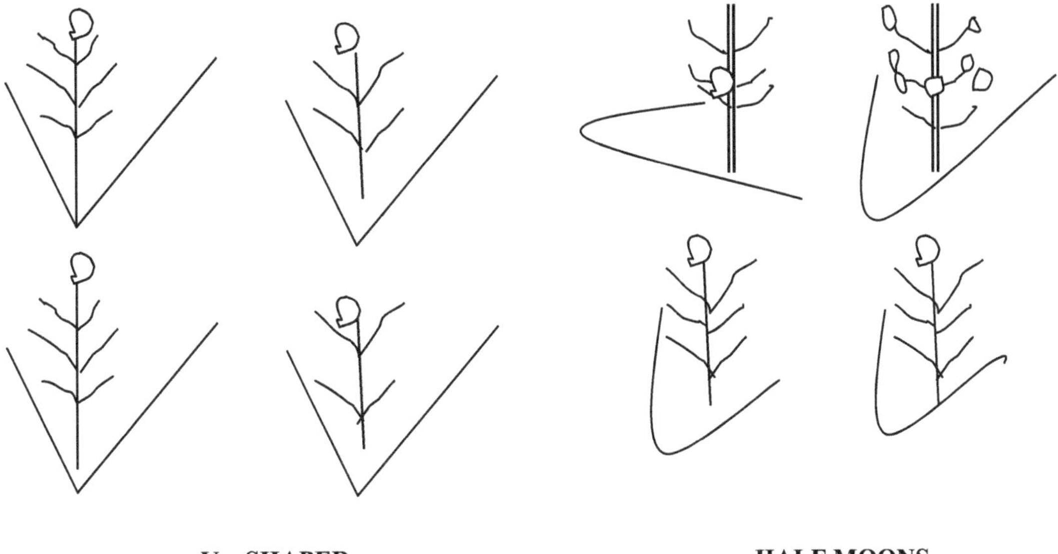

V – SHAPED   HALF MOONS

(c) *For sloppy lands (4-10 % slope)*: A series of ridges are constructed along contour lines or on gentle gradients to intercept and temporarily store surface water. Ridges of different shapes are possible like horse shoe, fishbone, gradonis etc. Ridges are made around individual tree to concentrate and retain run-off in the root zone.

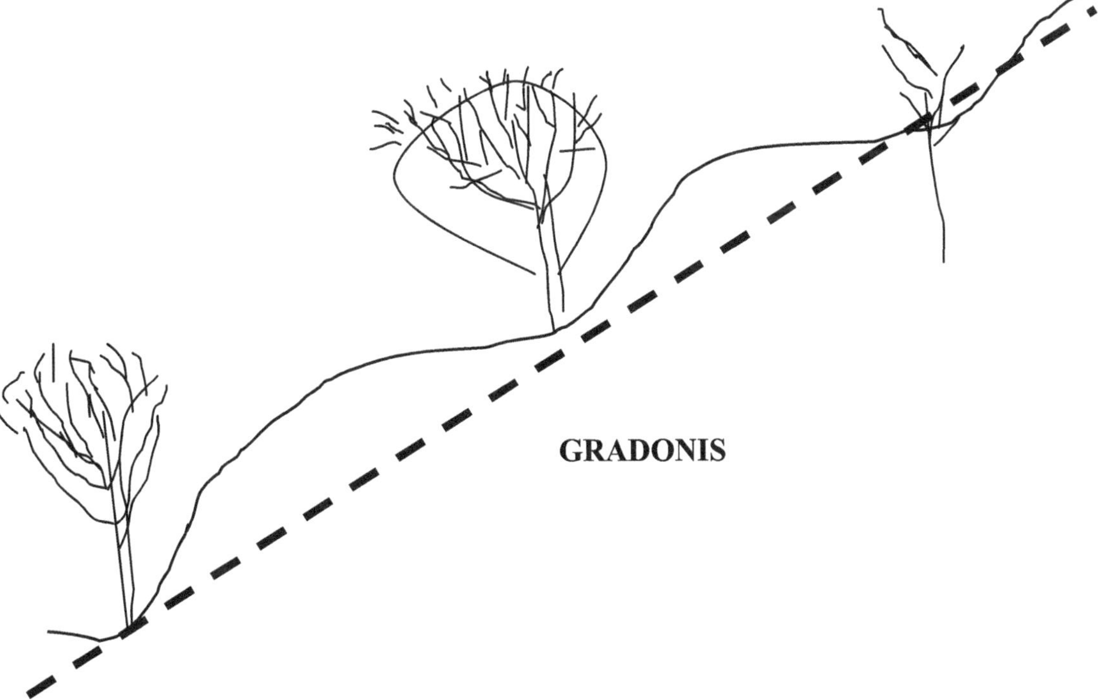

GRADONIS

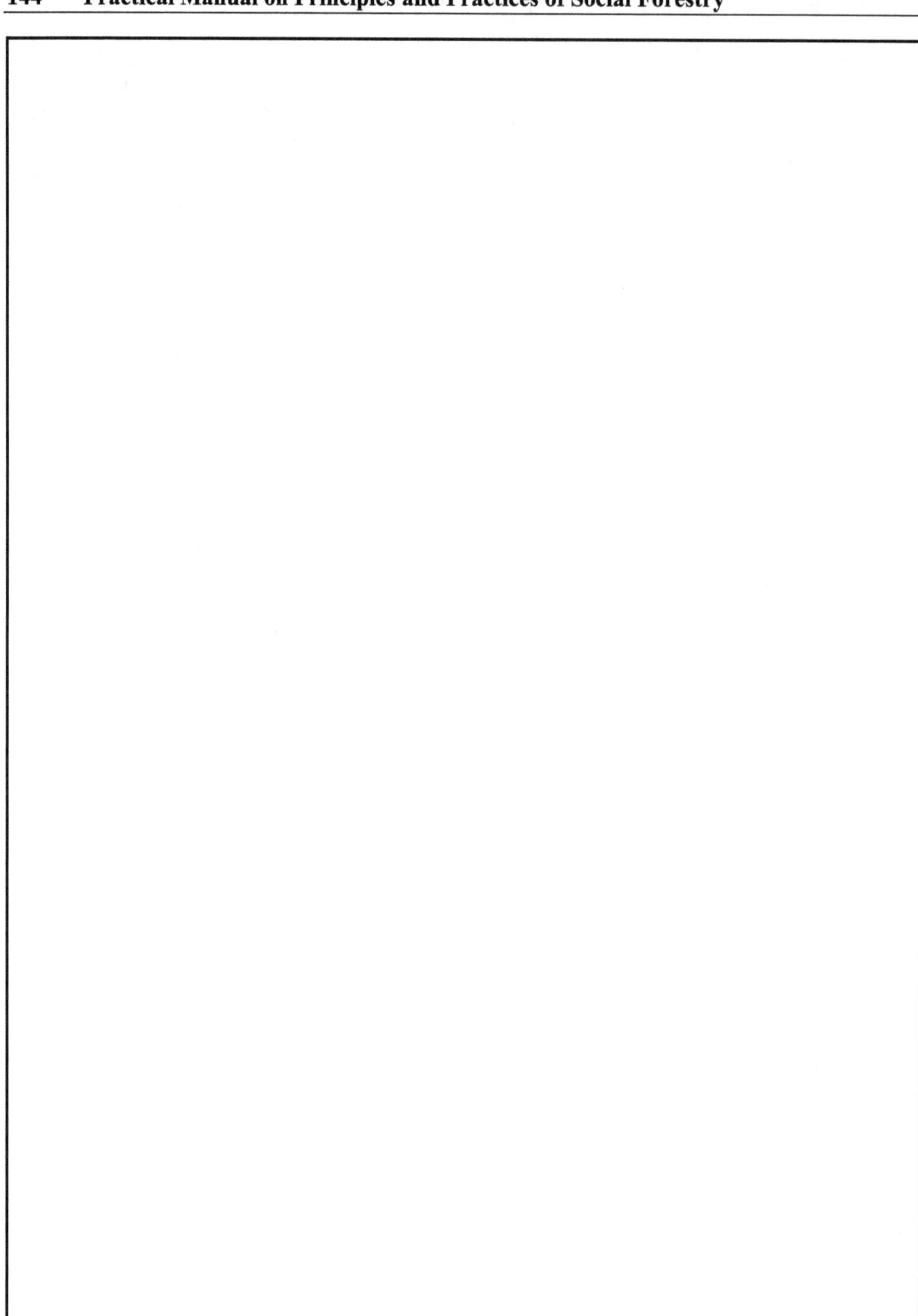

(d) *Compartmental mini-catchment (for slope beyond 10%)*: The field is divided into 16-100m^2 compartments by 0.3m^2 size bunds depending upon the spacing of trees. The tree is planted at lower corner of the compartment. The soil surface within the compartment is suitably modified to concentrate the run off in the specific area where the tree is to be planted.

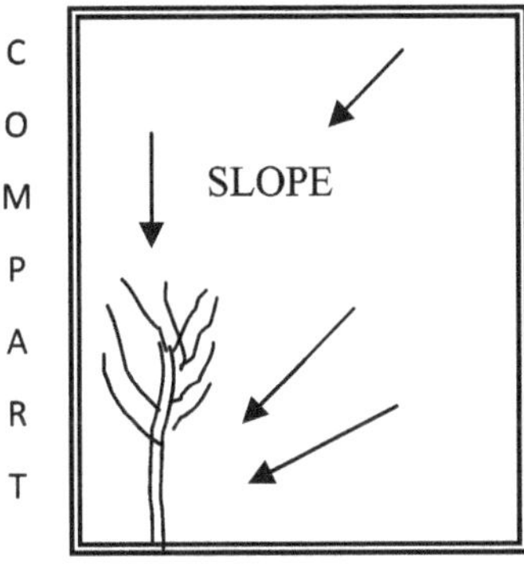

BUNDING

(e) *Contour furrow / trench planting system*: On hilly or undulating lands it is essential to plant the trees along the contours but across the slope upto a slope of 6%, contour furrow system will be more effective. Beyond that contour trench system will be more appropriate. These furrows and trenches with mild gradient will act as drainage ditches and excess run-off may be safely diverted to water harvesting ponds. Contour trenches will have an added advantage of prevention of soil loss on hilly terrains.

COUNTOUR ALIGNMENT ON HILLS

(f) *Double ring system*: This system is being largely practiced for the roadside plantation which require protection from the cattle. Two rings are made around the planted tree with a depth of 1m and 1-2m away from the tree so the inner ring will help in collecting rain water for moisture conservation.

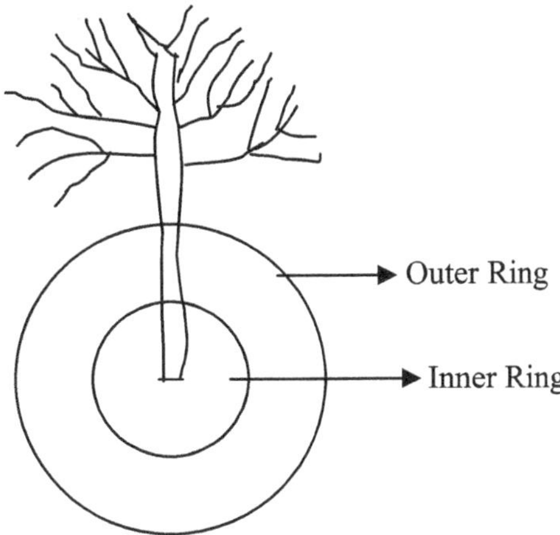

DOUBLE RING SYSTEM

Planting Techniques

1. ***Seed sowing***: Unless planting is the only successful method, seed sowing is the first mode of artificial regeneration in any area, mainly because of case in execution, low cost of labour and money. The time or season of sowing is very important factor affecting the success of artificial regeneration. It should be carried out in a planned way in early June just before the pre monsoon showers arrive. This leaves enough time for resowing, if any, which can be carried out after the first few showers.

2. ***Planting of saplings (methods)***

   (a) ***Planting with naked roots***: Plants are removed from the nursery only when the rains have fully set in and pit digging is complete. While removing seedlings from the nursery, care is taken to ensure that the roots are not damaged. Light bundles are made of the naked root plants so as to facilitate their quick transport to the forest.While planting, the seedlings are placed in a vertical position into the pit by holding it by the collar. After this the soil is pushed in from all sides, taking care that the roots remain in more or less vertical position.

   (b) ***Planting with ball of earth***: Many species are unable to bear the shock of naked root planting and hence may have to be transplanted with a ball of earth. The pits should be large enough to accommodate the entire ball of earth at planting.

   (c) ***Brick planting***: Plants grown in bricks are transported to planting sites by digging out the bricks form the nursery beds. The planting technique adopted is same as in case of ball planting. This method is followed in arid zones and sand dune areas and also coastal sites.

   (d) ***Container planting***: This is one of the most common methods of transplanting adopted in India. So far as transplanting is concerned, container plants may be grouped into two.

      (i) Those which are planted with container.

         e.g: donas, moss, fibre or betula bark container usually with light bottom. They may be buried into the soil, together with the roots.

      (ii) Those whose container have to be removed before planting.

         e.g: Earthen pots, polythene bags, metal container etc.

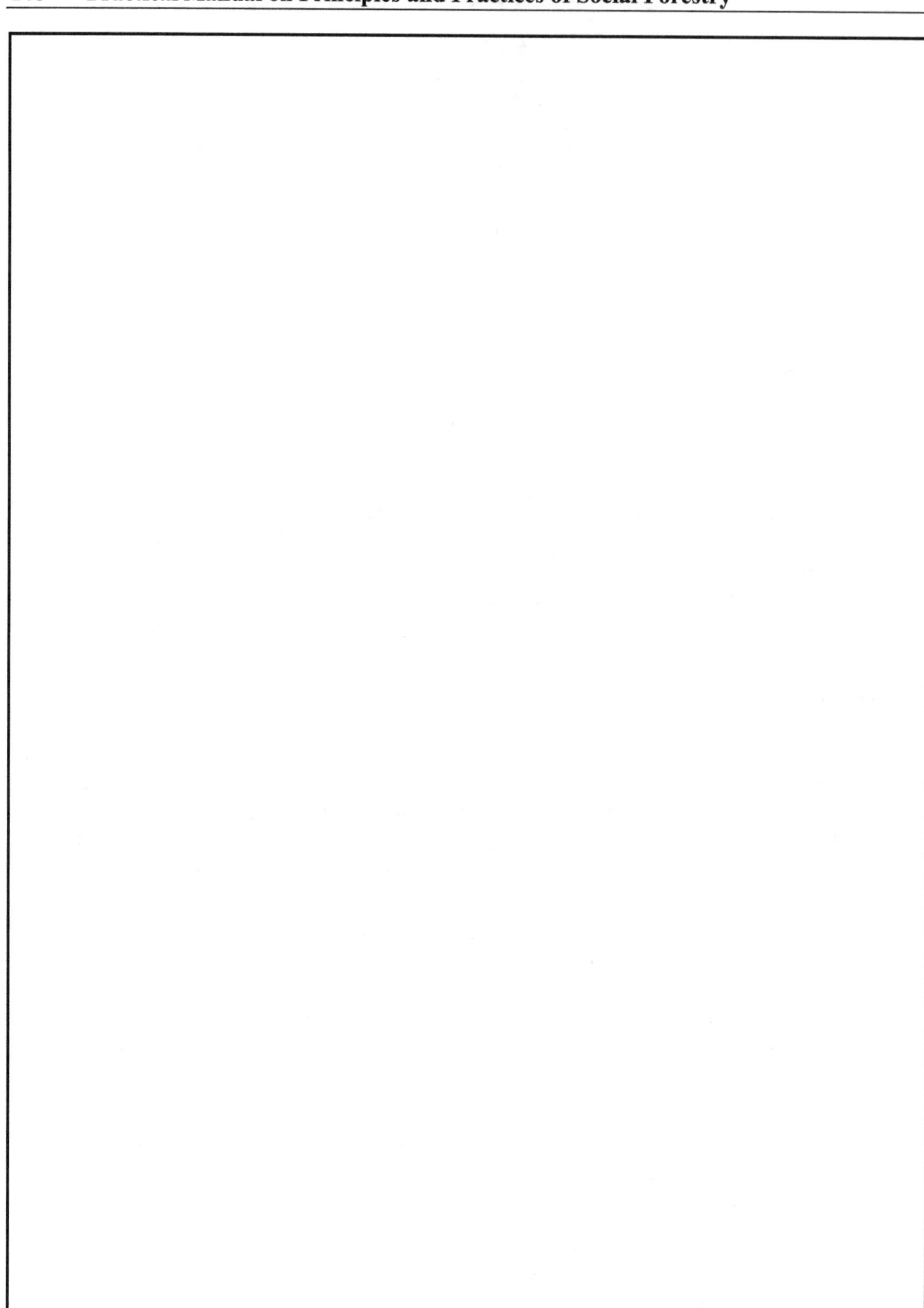

3. *Stump planting***:** This method of planting differs from the other methods as in it entire plant is not used. Stumps usually comprise of root shoot cuttings prepared and tied into bundles of 100 each and transported. Stumps are planted in the forest either by digging small pits or in crow bar holes. The stump is so placed that its collar and the shoot above it remains outside the soil.

Aftercare (Tending)

After planting, various types of tending operations are needed for promotion of growth at different stages of development of crop. Of these weeding and hoeing, cleaning and replacing, fertilizing, pruning and thinning and improvement fellings are considered important. On most plantation sites, tending is mostly concerned with preventing the plants from being suppressed by competing weed vegetation.

Weeding

Under normal conditions, it will be advisable to remove weeds immediately after the rainy season. Weeding may have to be carried out for upto the first 3 or 4 years of growth in case of slow growing species and perhaps for less number of years for fast growing species.

Objectives of weeding
1. To reduce root competition.
2. To improve light conditions in the forest, particularly if it has tall, dense weed growth.
3. To reduce the loss of water through transpiration.
4. To ensure a proper healthy growth and general well being of the seedlings.

Pattern of weeding
1. Complete weeding
2. Line weeding
3. Strip and inter-row cultivation
4. Spot ringed.

Method of weeding
1. Mechanical weeding
2. Hand weeding
3. Chemical weed control
4. Biological weed control

Cleaning

Cleaning is a silvicultural operation carried out in a sapling crop. It consists of the removal or topping of inferior growth including that of the favoured species, when they are interfering with the better individuals. As saplings pass on into pole stage, cleaning merge with thinning.
Cleaning involves the following operations.
1. Cutting back of shrubs and associated herbs which tend to interfere with the growth of saplings of the favoured species.

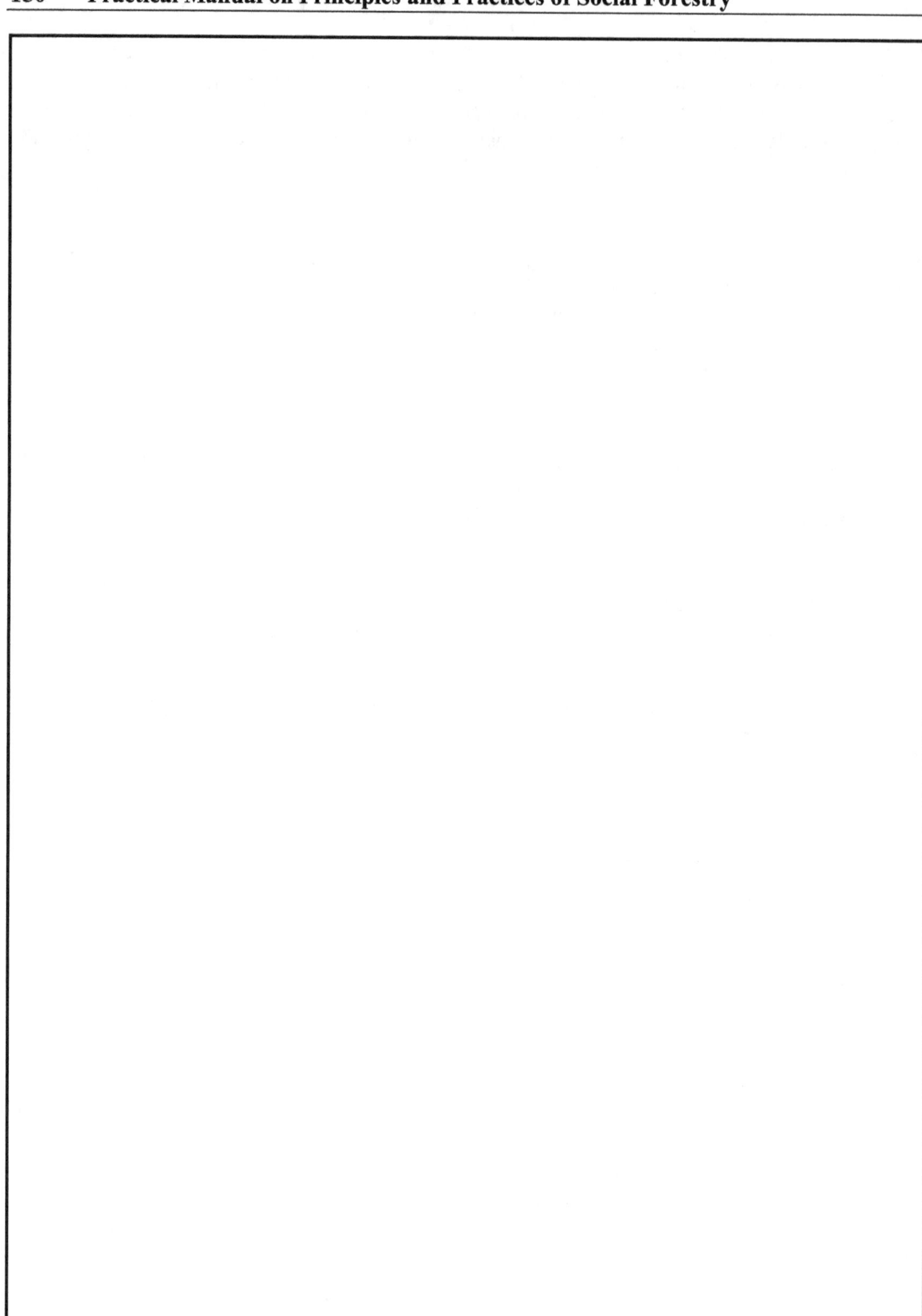

2. Cutting back of malformed, diseased or suppressed individuals of the favoured species.

3. Singling out of a flush of coppice shoots of the favoured species.

4. Climber cutting.

Respacing

It is removal of competing trees of the same species or favoured species under following situations.

1. When initially too close spacing has been adopted with the apprehension that few will establish.

2. When natural regeneration has come up after cleaning.

3. To thinning plants coming after direct seeding operations respacing is not advisable early on the dry sites of where grazing pressure are heavy or where climate is unpredictable. In other areas it may be completed within first two years of planting.

Pruning: It is an operation of removal of live or dead branches from standing trees for the improvement of the tree or its timber. Pruning is the instrumental technique in the hand of a forester to get knot free timber and improve wood quality. Generally the operations are distinguishable into two operations, low pruning and high pruning.

(a) *Low pruning (or) Brashing*: It is removal of branches upto a height of about 2m. up the stem just after canopy closure to make access to the stand to assess its vigour and to lay operations for fire protection etc.

In some tree species, natural death and fall of branches of standing trees due to deficiency of light, decay and snow occurs, it is called natural pruning. In other species pruning occurs naturally when the density of the crop is high.

(b) *High pruning*: It required when wide spacing and heavy thinning are practice. This is to get good quality timber.

Thinning: Thinning comprises of fellings or removal made for better form of the tree that remain without making a permanent opening in the canopy. Thinning is also carried out to provide intermediate financial return from sale of thinning material. Thinning starts 5 years after planting and is carried out at an interval of 5 years till stand matures.

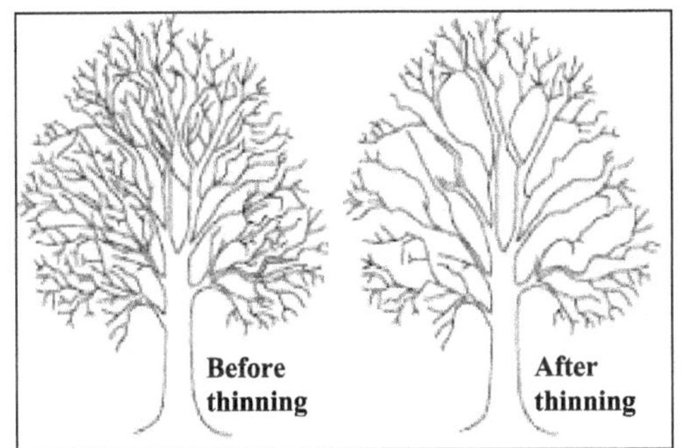

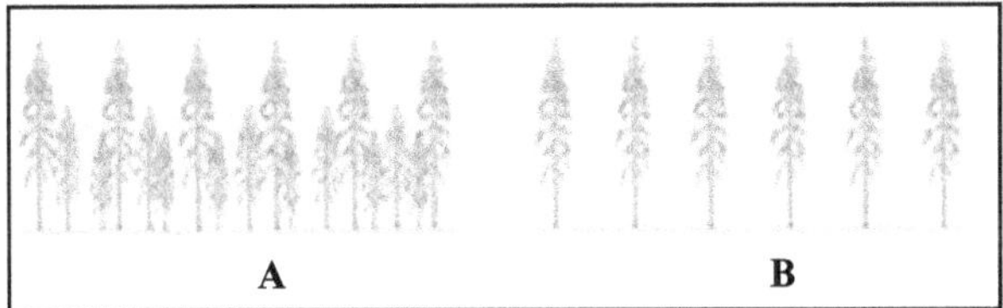

Thinning for forest health heavy thinning
A. Light thinning B. Heavy thinning

The various thinning methods are :

1. Mechanical thinning
2. Low thinning or ordinary thinning
3. Crown thinning
4. Free thinning
5. Advance thinning
6. Maximum thinning.

Improvement Felling: It is the removal of less valuable trees in a crop in the interest of better growth of the more valuable individuals. Usually applied to a mixed, uneven aged forest.

***Mulching*:** Mulching is application of any plant residue and other materials used as a covering for the soil to conserve moisture, reduce run off and erosion, check weed growth, protect from winter climate or improve the soil.

Straw mulching

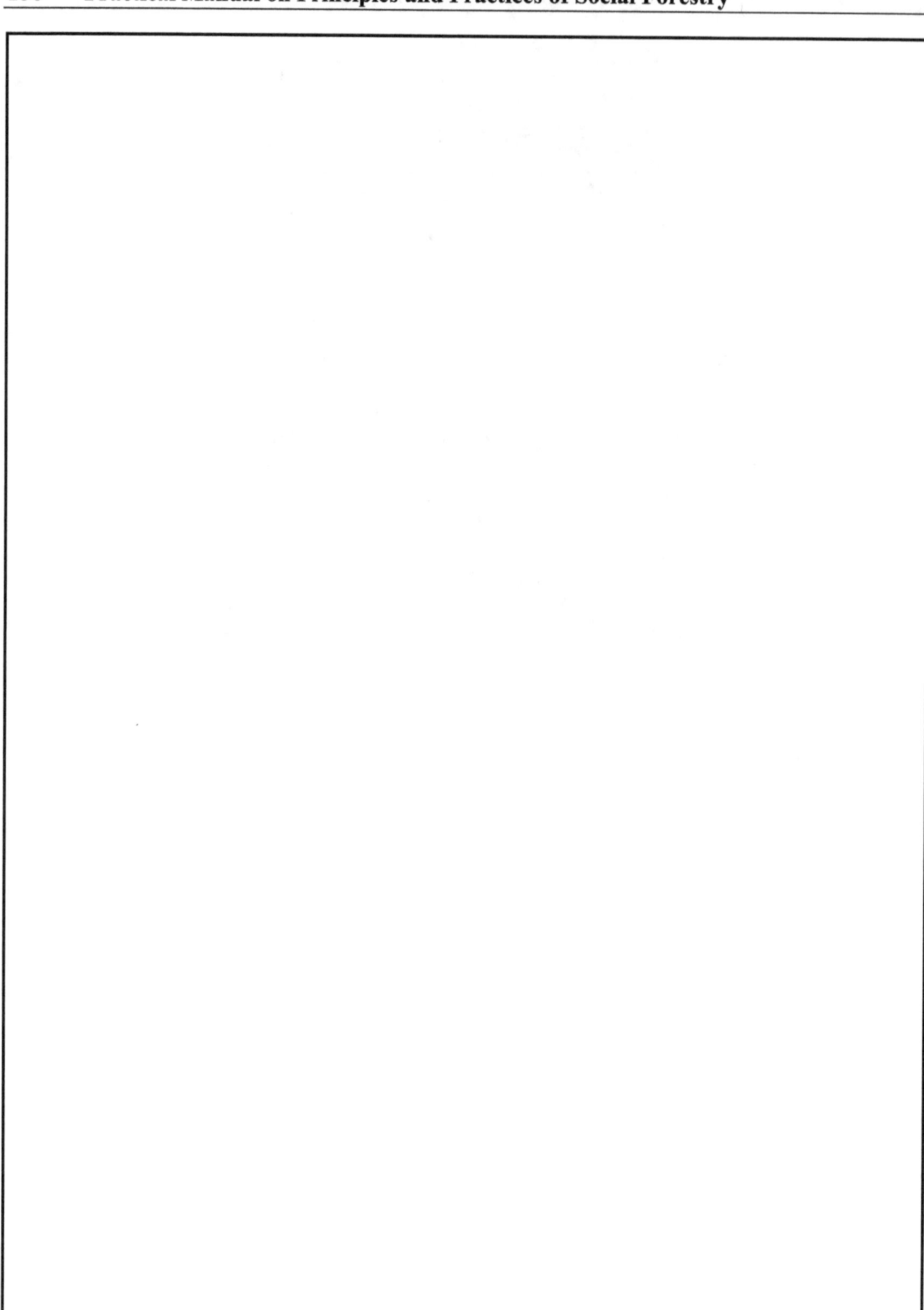

BIOMASS ESTIMATION IN ENERGY PLANTATIONS AND TREE HEIGHT MEASUREMENT

Objectives:

1. ____

2. ____

3. ____

Biomass estimations are essential for estimating net primary production (NPP), understanding nutrient cycling, energy transfer, predicting the effect of tree utilization management procedures and stability of forest stands.

For biomass estimation, two distinctly separate situations can be considered

1. Destructive sampling and weighing.

2. Estimation by prediction equation based on samplings.

I. **Destructive sampling and weighing:** The steps followed in destructive sampling and weighing are:

    (a) *Selection of a sample plot*: Generally it is not possible to measure all the trees of the forest stand and thus sample plots are taken to represent the forest.

    (b) *Sample tree parameters*: There is always a high correlation between DBH (Diameter at breast height) and height and weight. Thus it becomes essential to measure the DBH of all trees in a sample plot and height of some trees. Sample trees should be measured before felling in an identical way to those trees within the sample plot for which weights are to be estimated.

All the tree components viz., foliage, twigs, branches and bole are separated immediately after felling and their fresh weights are recorded in the field. A representative sample of each tree component (100 g for foliage, twig and branch) is taken for its oven dry weight estimation at 80^0 C. The bole portion of the sample tree is cut into 100 cm or 1 m long sections for the convenience of weighing. A 5cm wide disc is removed from the base of each billet for the estimation of fresh and oven dry weights of bark and wood (under bark) and also for the estimation of volume (over bark and under bark) of the main bole (upto a diameter limit of 5 cm over bark).

Estimation of stand weight from samples trees

There are several general categories of methods used to estimate the stand weight from sample tree data.

    (i) *Mean tree technique*: This method is based on the detailed measurement of one or more trees of mean dimensions. Sample trees data are converted to a land area basis by multiply the biomass of different components by the total number of tree per unit area.

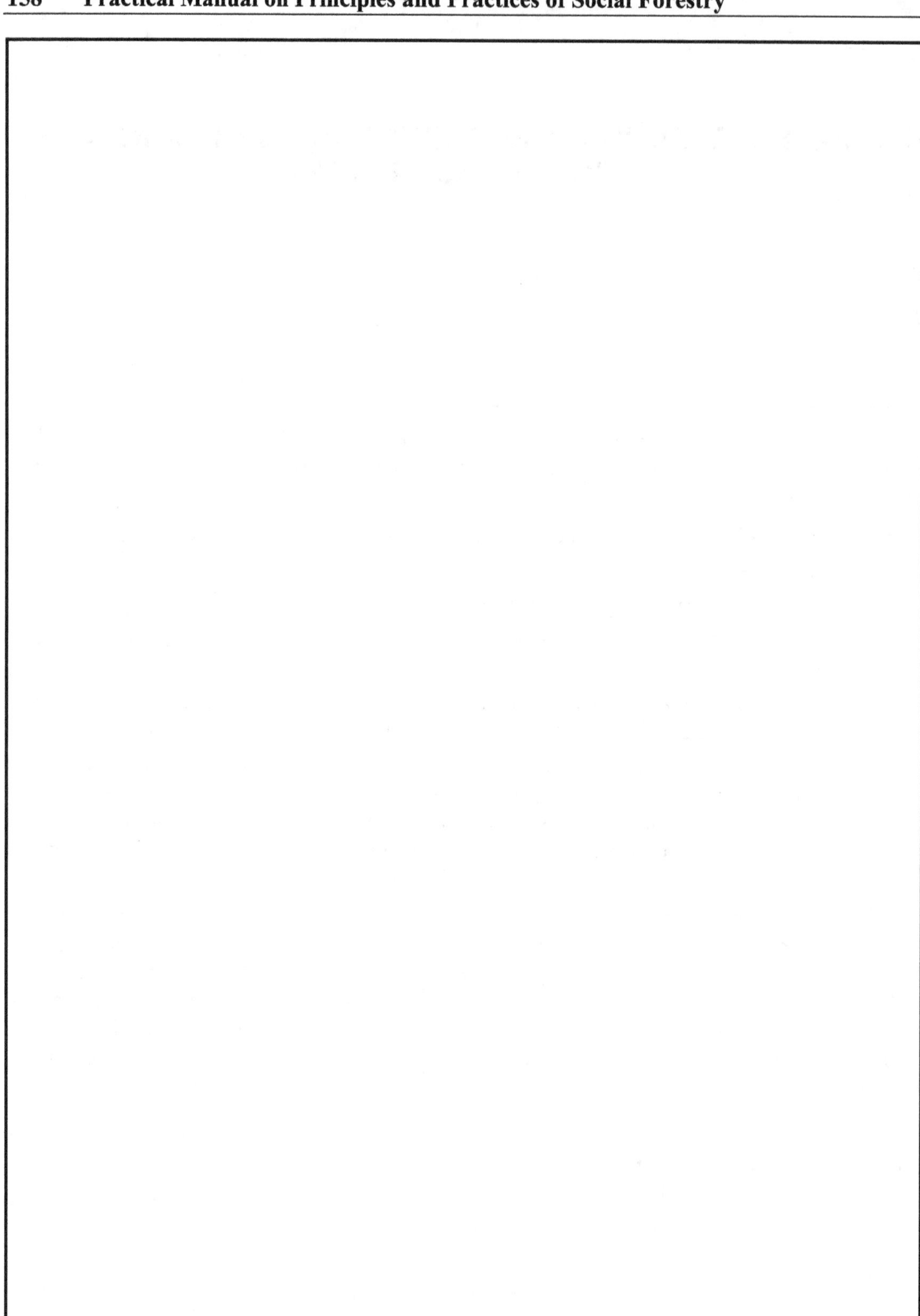

(ii) *Stratified tree technique*: This method is almost similar to the mean tree technique. The stand is stratified by different diameter classes. One or more trees are harvested from each diameter class having parameters close to the mean of the class and biomass figures of different components of each stratum and added up to get the biomass on unit area basis.

(iii) *Unit area method*: This method involves the harvesting of all the trees standing on a randomly selected plot. But this method is too much time consuming and seldom used in biomass studies.

(iv) *Basal area proportion method*: In this method a tree of mean basal area is harvested and weight multiplied by the number of trees per unit area gives biomass. When a tree of mean basal area is not available, several trees close to the mean basal area are sampled and the following formula is used for biomass estimation.

$$W = w \frac{G}{g}$$

where

W = Weight of trees on the plot.

w = Sum of the weights of sample trees.

G = Sum of basal area of the plot

g = Sum of basal area of the sample trees.

A better estimate is obtained if stratification is done by the diameter and taking trees of mean basal area stratum.

II.   Regression estimation technique

This method is also called "Dimension analysis of woody plot and algometry".

This technique involves the following three steps.

(i)   Trees are tallied by species and DBH.

(ii)   The weight of the sample trees are related to easily measured parameters (DBH, height etc.)

(iii)   Application of regression obtained in step 2 to step 1.

VOLUME ESTIMATION

Forest yields are generally measured in volume i.e., the volume over bark of stems and branches exceeding 7 cm diameter. The yield may be (a) standing i.e., what is on the hoof now (b) final i.e., what would be obtained by clear felling at a selected age (c) "Intermediate or thinning" i.e., what can be or has been removed in thinning d) "Mean annual" i.e., sum of all yields including the standing, divided by the age. It is commonly known as mean annual increment (MAI).

Yields of volume are a function of diameter and height and number of trees and in case of MAI of time of age as well. The formulae for calculating of volumes are as follows.

(a)   **For single tree**

$$V = g \times h \times f$$

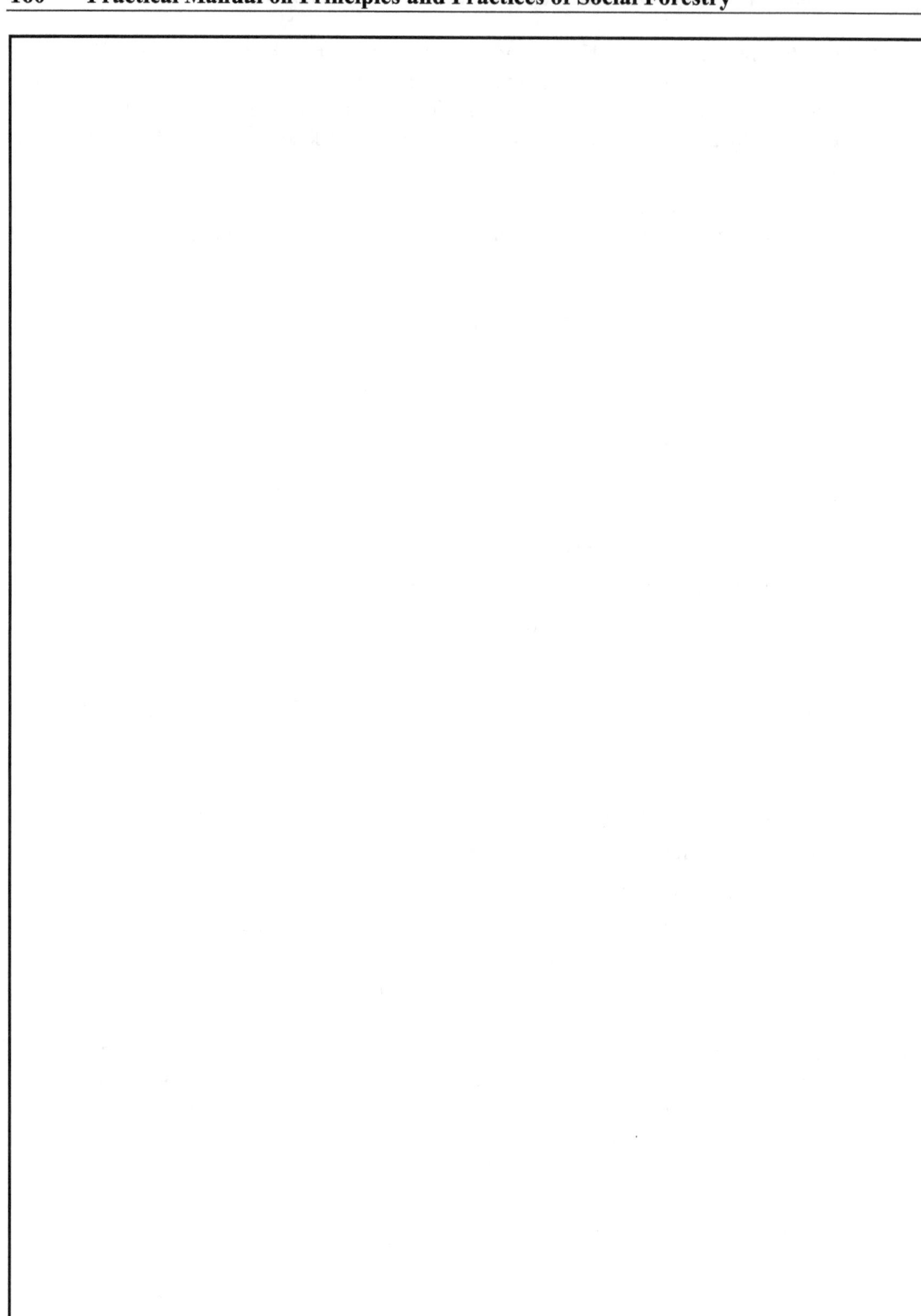

where

  V = Volume over bark.

  g = Basal area i.e., $\dfrac{(m^3)}{4}$ at breast height

where

  d = diameter over bark at breast height i.e., 1.3 m height from ground level

  h = Total height from ground to tip (m)

  F = Form factor which is the ratio of actual volume to cylindrical volume (0.7)

  'V' = cylindrical volume.

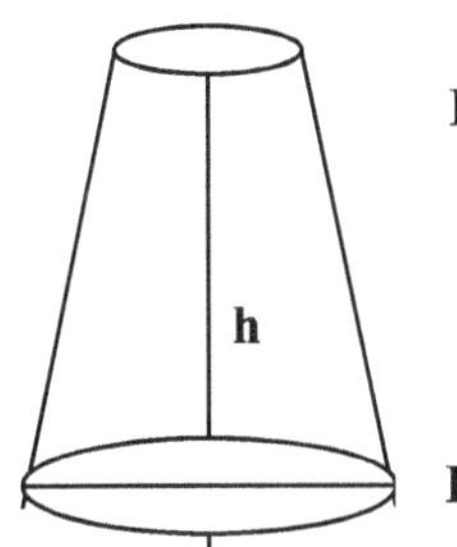

$$\text{Actual volume} = \pi \frac{(D_1 + D_2)^2}{4 \times 2} \times h - 'A'$$

$$\text{Cylindrical volume} = \pi \frac{(D_1)^2}{4} \times h - 'B'$$

$$F = \frac{B}{A}$$

(b) For stands of trees

$$V = \frac{N \times G(h - f)}{\text{Plot area}}$$

where

  V = Volume of stand

  N = Number of stems per hectare

  G = Total basal area per hectare

  h = Dominant height, defined as the mean of the 100 thickest trees on a hectare.

  f = Form factor

Trees volume tables: A volume table is a tabulation that provides the average contents for standing trees of various sizes and species. The principal objective in completing such tables is to obtain an estimate for standing trees that would correspond with the volume obtained if the same trees are felled. Thus such tables are used in timber estimating as a means of ascertaining the volume and value of standing trees in a forested tract.

The volume tables that are based on the single variable of DBH are commonly referred to as "local volume tables", those that required tree heights and form or taper are referred as "Standard volume tables".

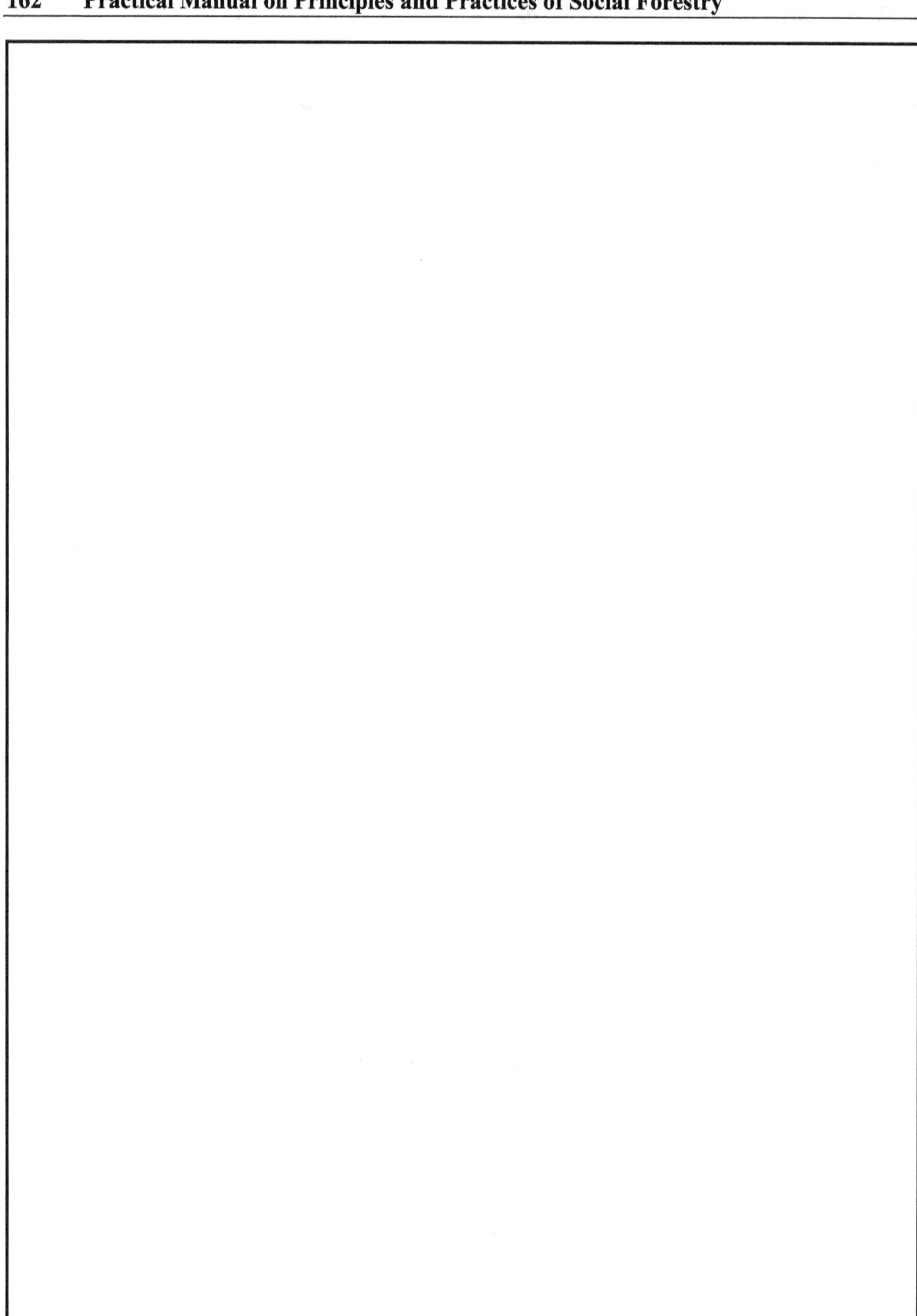

Tree height measurement with the help of Altimeter

Description: The altimeter of multimeter is made up of aluminum, moving parts being enclosed to withstand shock or damage. It measures height of objects, angle of elevation in degrees and the corrections for heights in sloping or hilly regions. It is compact and light weight and weights only 350 grams.

Trees height measurement in plain grounds

Procedure: Press the lever (L) and see that the pointer is moving freely. Hold the instrument firmly. The tree, whose height is to be measured is than viewed through Eyecup (EC) and target (T) such that the target is in line with the top of the tree. Then press gently the bottom (M). This will arrest the pointer. If the observations are being made from a distance of 20 m, than the scale A is read off, if the relevant distance is 30 m, then the scale B is read off. The pointer reading in both cases gives the height of the tree (h_1) above eye height of observer. In order to get the total height of the tree, the eye height of the observer (h_2) is added to the pointer reading (h_1). If the observations are being made at a distance other than 20 m or 30 m, than the scale E is read off. The percentage reading of the pointer is multiplied with the horizontal distance from the tree to observer. It gives the height of tree above the eye height of the observer. By adding the eye height of the observer the total height of the tree can be got.

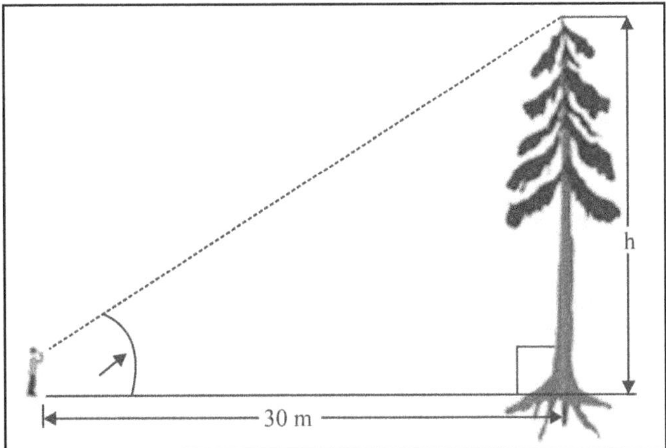

Addition of eye height can be avoided by the following procedures.

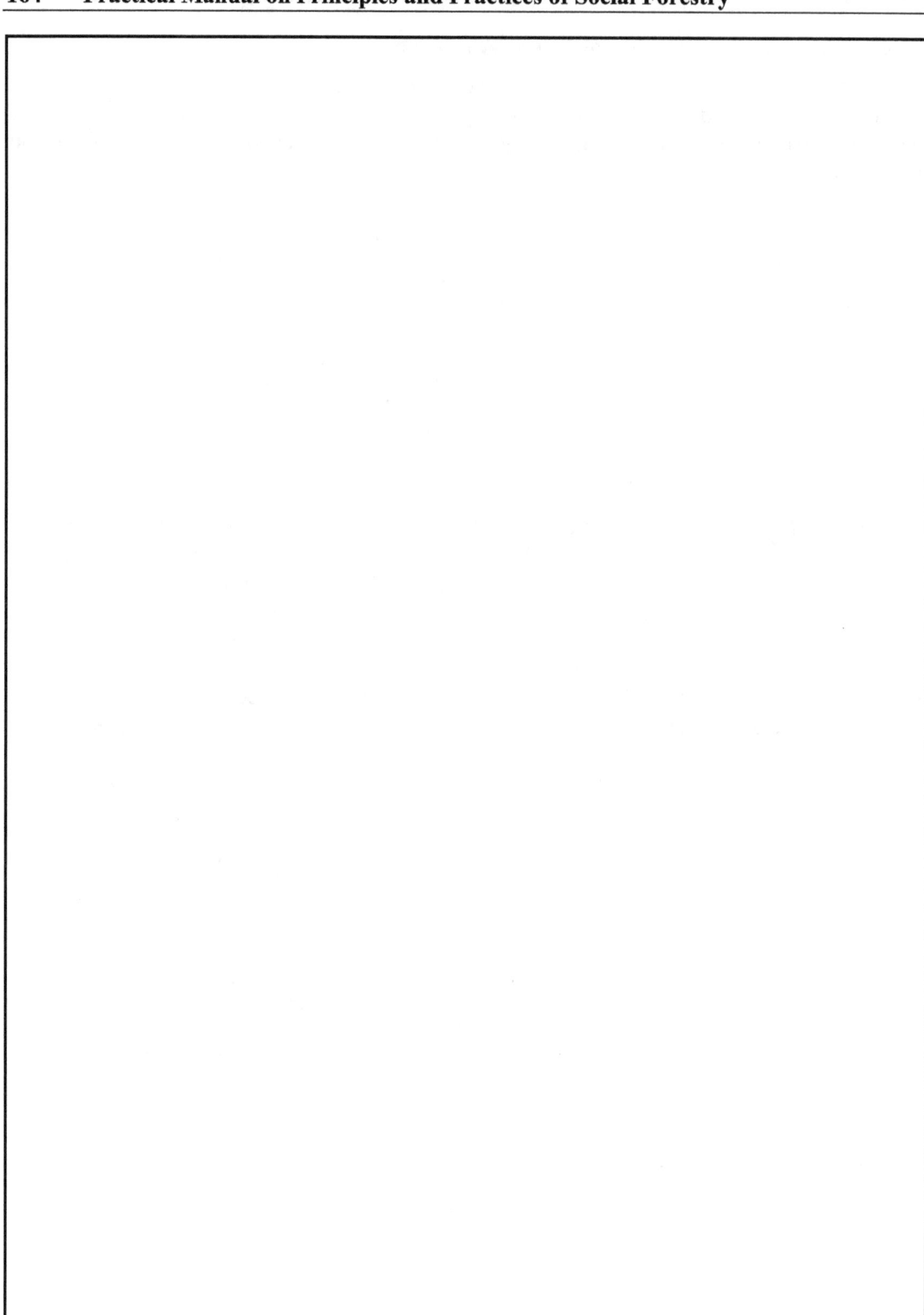

First take the reading of the top of the tree as per the procedure given above (h_1). Then take the reading of the base of tree. For this, release the pointer and view the base of the tree such that the base is inline with the target when seen through the eyecup. Gently arrest the pointer. Then read off the pointer reading, selecting the scale (A, B or E) according to the baseline used. This will give the eye height of the observe (h_2). Then the two reading (h_1 and h_2) are to be added and the sum got is the height of the tree.

Tree height measurement in sloping or hilly regions

Procedure: In the sloping ground, the base line measured along the ground is not the horizontal distance between the tree and the observer. Then the correction of base line is necessary.

The tree whose height is to be measured is first marked. The mark is made at eye-height of the observer. Now form a suitable distance, this mark is viewed through eyecup and target. The pointer is arrested and the scale 'D' is read off. If the pointer reading is 100% then the ground is leveled and there is no need for any correction. If the pointer reading is other than 100% than the correction is necessary. In that case, the pointer reading of scale D is multiplied with the sloping distance between the tree and the observer. The product is the correct based line.

Then the top and the base of the tree are viewed and the pointer reading of scale E are noted. The two readings are added if these are on opposite side of zero and subtracted if these are on same side of zero. The total percentage thus got is multiplied with the corrected base line got earlier and the result is the correct height of the tree.

Measurement of angle of elevation of the tree

The angle of elevation of any object (tree) can be measured in degrees by this instrument. The object or tree whose angle of elevation is to be measured is viewed through eyecup and target. The pointer is then gently arrested. The pointer reading of scale 'C' gives the angle of elevation subtended by the object at point of observation.

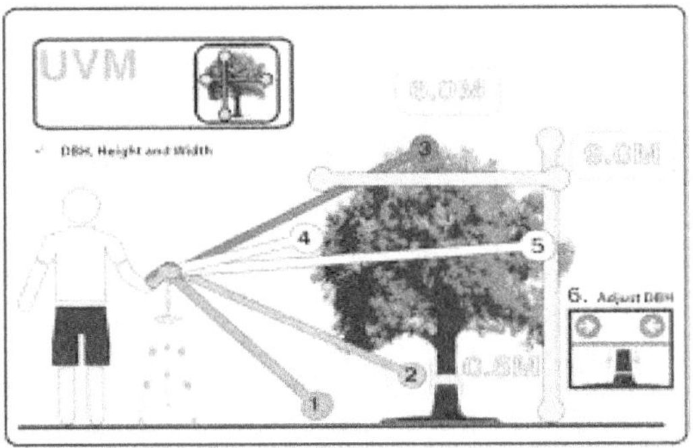

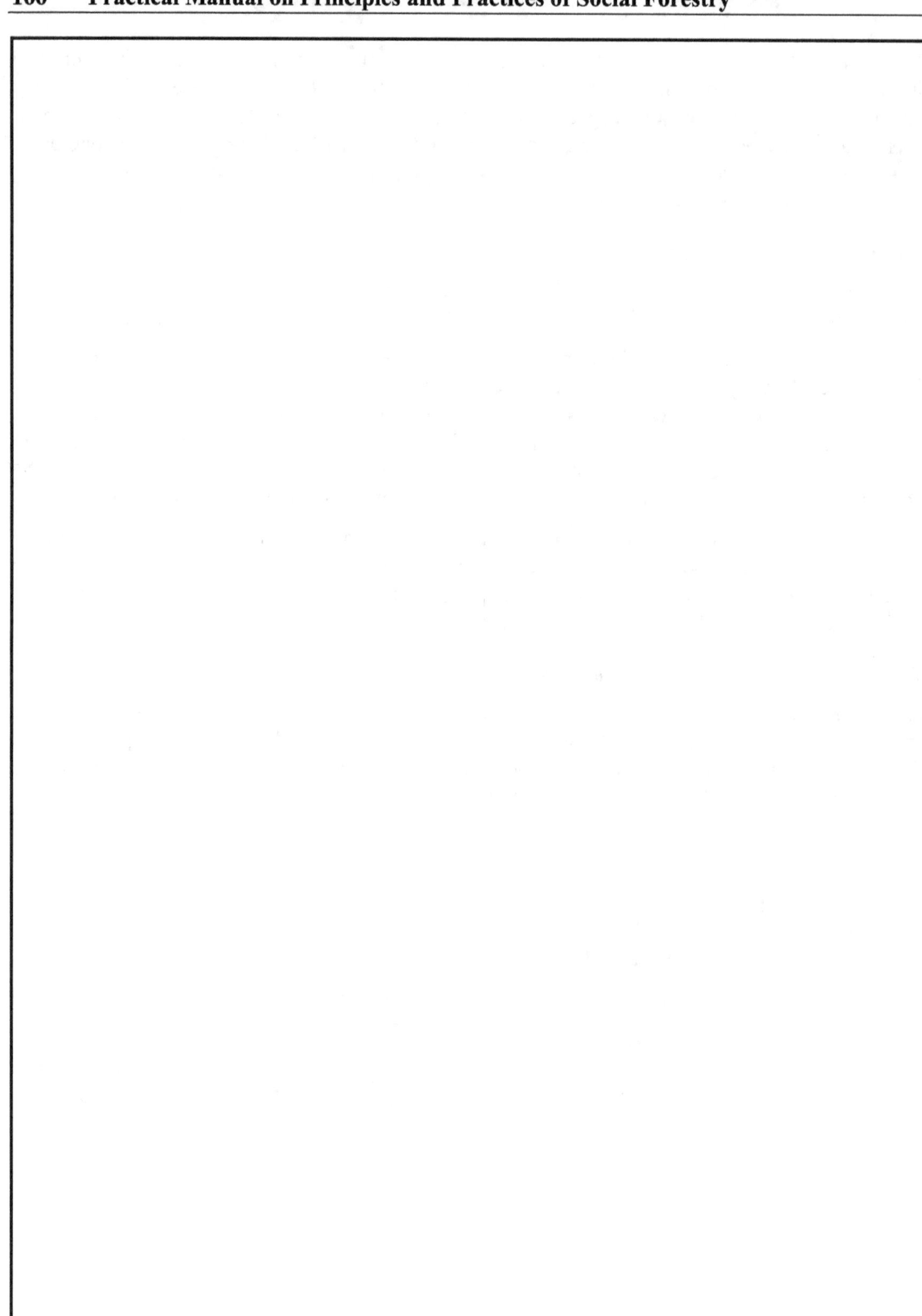

FIELD EVALUATION OF DIFFERENT AGROFORESTRY SYSTEMS

Objectives:

1. ____

2. ____

3. ____

Agroforestry is a sustainable land management system which increase yield of land, combines crops, woody perennials and/or livestock together simultaneously or sequentially on the same piece of land and applies management practices that are compatible with the cultural practices of local population.

Based on the component combinations the Agroforestry systems are classified as follows:

1. *Agri-silviculture*: In this system integration of forest trees with arable crops is practiced mostly on the arable lands to get both food and wood. Nitrogen fixing trees (NFTs) are preferred which supplement the nitrogen requirements of crops through leaf fall, lopping or green leaf manuring. In semi-arid tropics where residual build up of organic matter is very low and in drylands where resource poor farmers cannot afford to apply costly nitrogen fertilizers. The NFTs are of great help.

2. *Alley cropping*: It is a type of agri-silvicultural system where in food crops are grown in alleys formed by hedge rows of trees or shrubs (the space between two hedge rows of trees is called alley). The essential feature of this system is that the hedge rows are cut back and kept pruned during cropping season to prevent shading, to reduce the competition with food crops. This system is highly suitable for marginal lands of arid regions.

Advantages

1. It provides higher total biomass per unit area.
2. It provides green fodder during lean period.
3. It utilizes the off-season precipitation effectively.
4. It provides additional employment during off seasons.
5. Reduces soil and water erosion when planted along the contours on sloppy lands.
6. Provides green manure to component food crops.
7. Pruned material can be applied as mulch.
8. Reduces soil temperatures, enhances activity of soil microbes.
9. Provides biologically fixed nitrogen to associated crops.
10. Improves the income of farmers.

   Ex: Subabul, *Acacia albida* are highly suitable tree species for alley cropping.

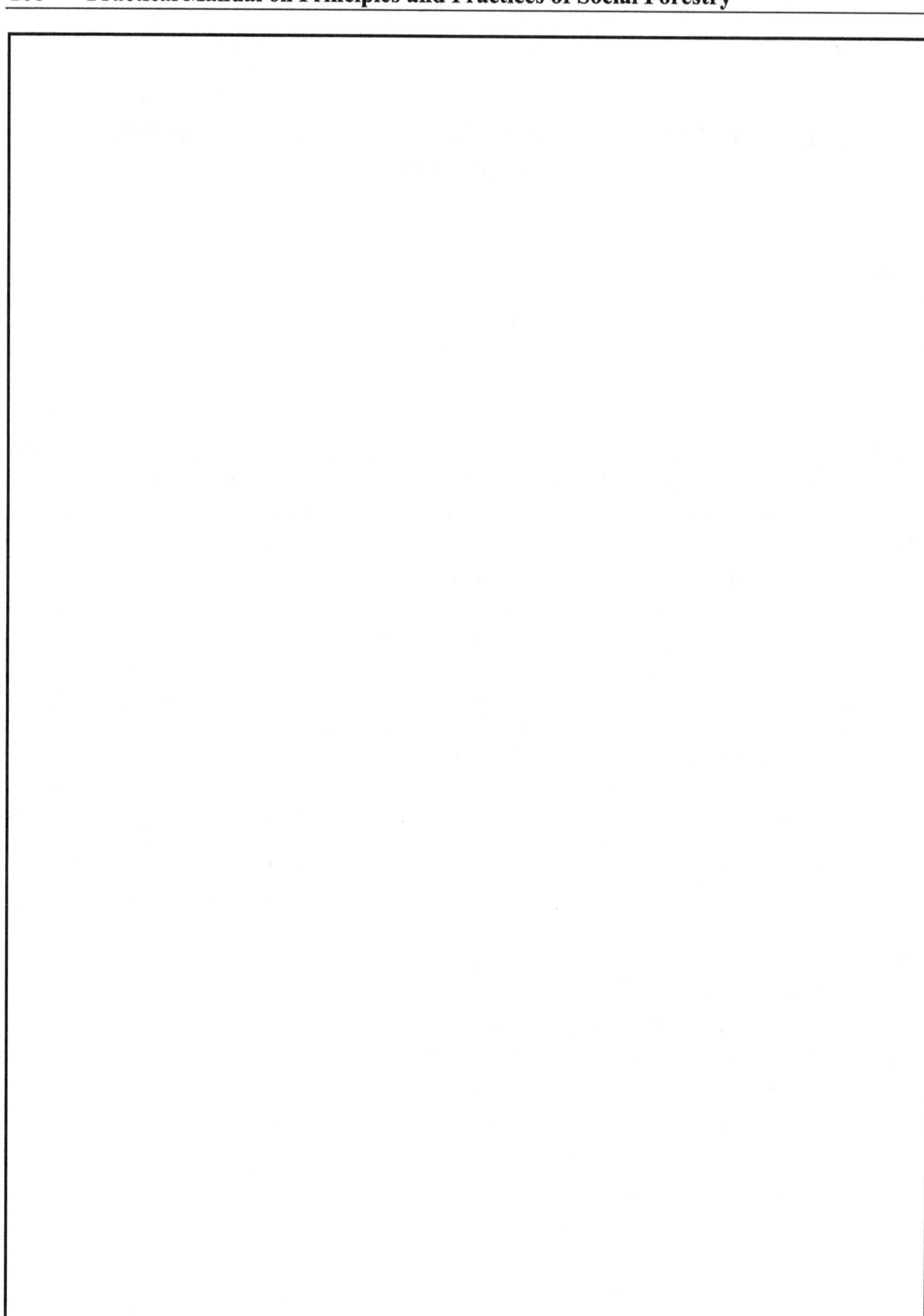

3. *Agri-horticultural system*: It is a system of integration of fruit trees with food crops. Most of the fruit plants develop full canopy after several years and some of them require regular pruning thus permitting an intercrop. Intercropping in orchards keeps the weeds checked and the yield of intercrops will be a bonus to the farmer. Ber, amla, custard apple, phalsa, pomegranate and to some extent guava are suitable fruit trees under this system.

4. *Silivi-pastoral system*: It is a two-tier model of fodder based agroforestry system and is suitable for class IV and above lands. This system integrates top food, tree species with under storey forage crops to support livestock and to prevent land degradation.

Advantages

1. Provides fodder from under storey forage crops and top tree in fodder scarcity areas.
2. Improves soil nitrogen in poor shallow soils if legume pastures and trees are grown.
3. In areas of low and erratic rainfall where crop failure are common this system is better suited than agri-siliviculture.

Consideration in selection of tree component

1. Fast growing, having multipurpose usages i.e. for fuel, fodder, small timber etc.
2. High palatability and digestibility of foliage.
3. Good coppicing ability.
4. Ability to withstand browsing, trampling and intense lopping.
5. Resistance / tolerance to drought and extreme temperatures.
    Examples: Subabul, Sesbania, Ailanthus, etc.

In selection of fodder component

1. It should be able to grow well as under storey and compatible with other forage crops.
2. Must be prolific seeder or should be able to propagate vegetatively.
3. Possess high palatability and good nutritive value.
4. It should be able to withstand overgrazing and trampling
    Suitable pastuer: Stylo, Serrato, Cenchrus etc.
5. *Horti-pastoral system*: This is similar to silvipasture system except that the tree component is a slow growing fruit tree like tamarind, jamun, aonla, bael, cashew, wood apple etc. or fast growing fruit trees like ber, pomegranate etc. The agri-horticulture system in the advance stage of tree growth after canopy closure can be converted to horticultural system as the pasture grasses come up well under shade than other intercrops.
6. *Timfib (Timber-cum-fibre system)*: This is one of the agroforestry system mostly suitable for non-arable lands i.e. degraded sites. It involves integration of some timber yielding tree species with fibre yielding plants like sisal (*Agave* spp.). This was proved more remunerative than arable cropping when subabul was intercropped with sisal at Dry farming Research Station, Bijapur, Karnataka.
7. *Agri-silivi pastoral system*: In this system the trees are integrated with grasses and cattle are allowed to graze. Sometimes arable crops are grown as under storey crops with the trees and grasses like panicum and elephant grass are grown along the rows of trees which serves as vegetative barrier to control the erosion.

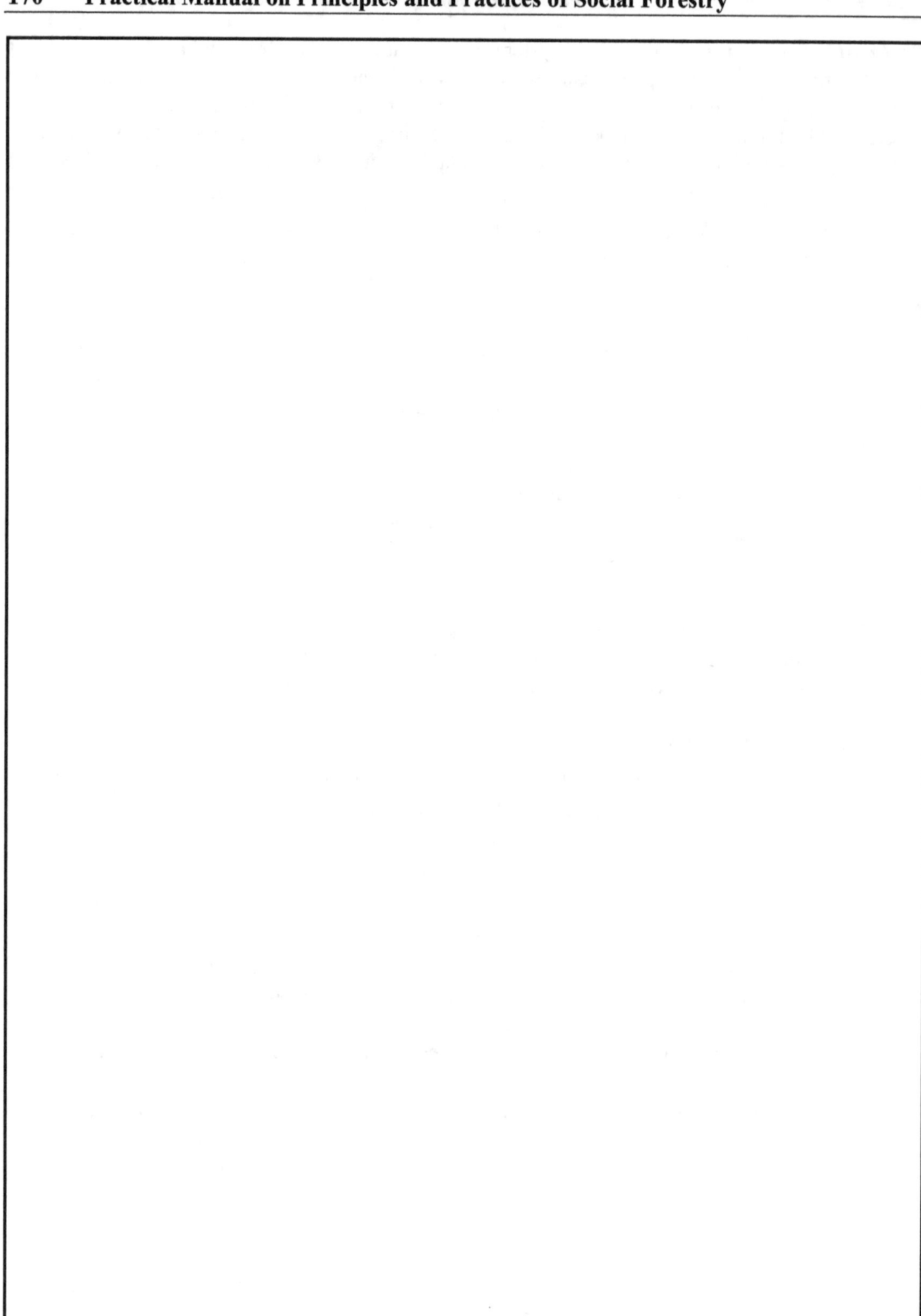

8. ***Multistoreyed cropping system:*** Multistoreyed cropping is highly suitable for small holdings in areas receiving higher rainfall (1200 mm/yr). High-density multispecies systems are capable of generating high biomass, more income and can meet various needs of the farmer.

In coconut, arecanut plantations, annual crops such as tapioca, elephant foot yam, dioscorea, turmeric, ginger, colocasia. Sweet potato, groundnut, vegetables and pulses, semi-perennials like pineapple, banana, papaya and perennials such as cocao, pepper, coffee, clove, fruit, fuel and timber trees can be grown.

Suitability of different Agroforestry systems for different soils

A. For marginal lands (class IV and above lands)

1. Silvi pastoral 2. Horti-pastoral, 3. TIMFIB.

B. For arable lands (class I, II and III)

1. Alley cropping
2. Agri-horticulture
3. Agri-silviculture (with NFTs)
4. Agri-silvipasture
5. Multi-storeyed cropping.

Disadvantages of agroforestry system

1. Prolific seeding of certain top feed trees resulting in weedy growth and depressing the growth of arable crops eg:- Subabul.
2. Root and shade effect of trees on cultivated crops. In case of *Acacia nilotica* this effects is more pronounced even upto or beyond 20 meters from the tree.
3. Tree horbours, birds, pests and diseases like *Leucaena psyllids* (*Heteropsylla cubana*) in near future may become pest of cultivated crops.
4. Agroforestry systems are labour intensive and use the scope of farm mechanization.
5. Allelopathic effects of trees in crops   eg :- Eucalyptus spp.

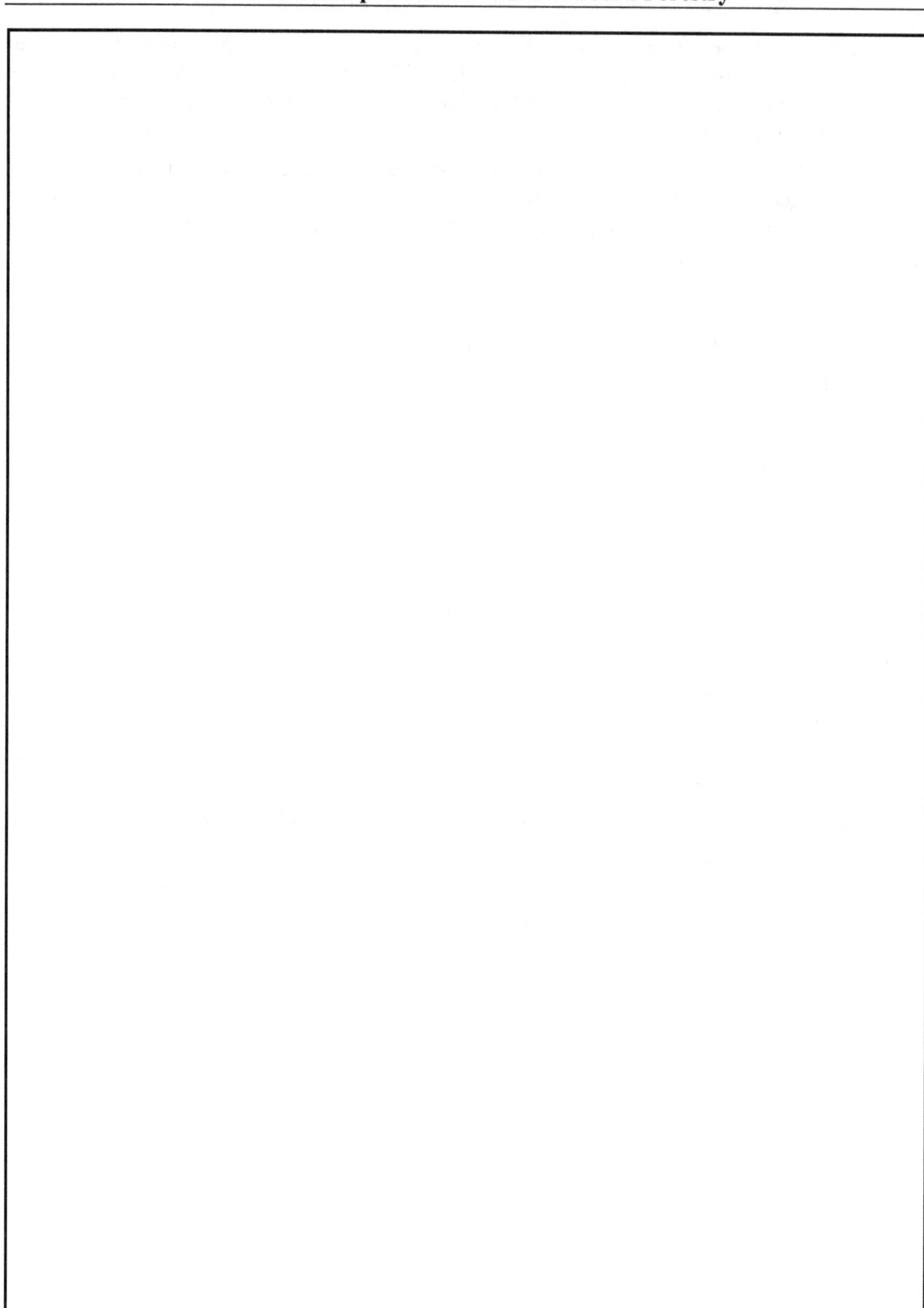

FIELD EVALUATION OF SHELTERBELTS & WINDBREAKS

Objectives:

1. ____

2. ____

3. ____

Design Criteria: The term windbreak refers to one or two rows of protective planting of trees where as shelterbelts is a more extensive and a long barrier. Windbreaks are planted perpendicular to the direction of wind in a given locality.

Generally in our country the windbreaks are planted in North-South direction to meet the wind forces from Southwest and Northeast.

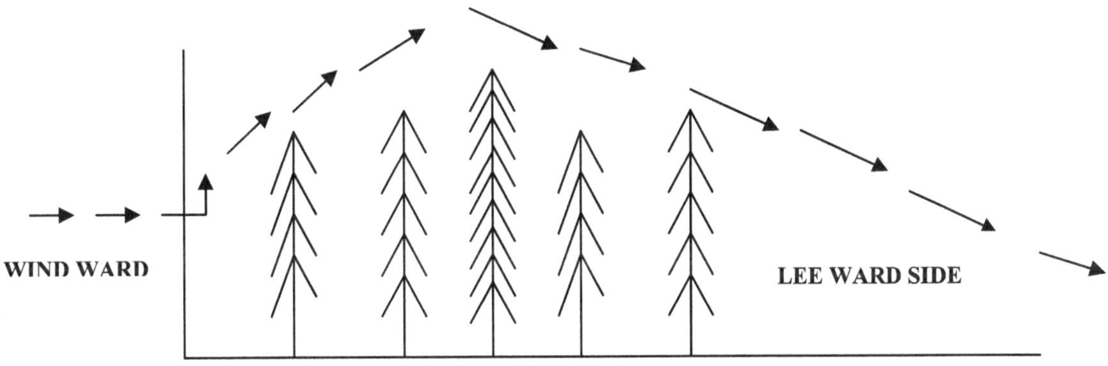

DESING OF WINDBREAK

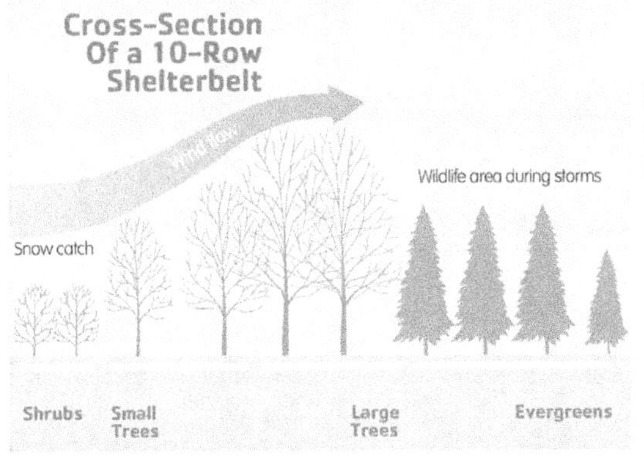

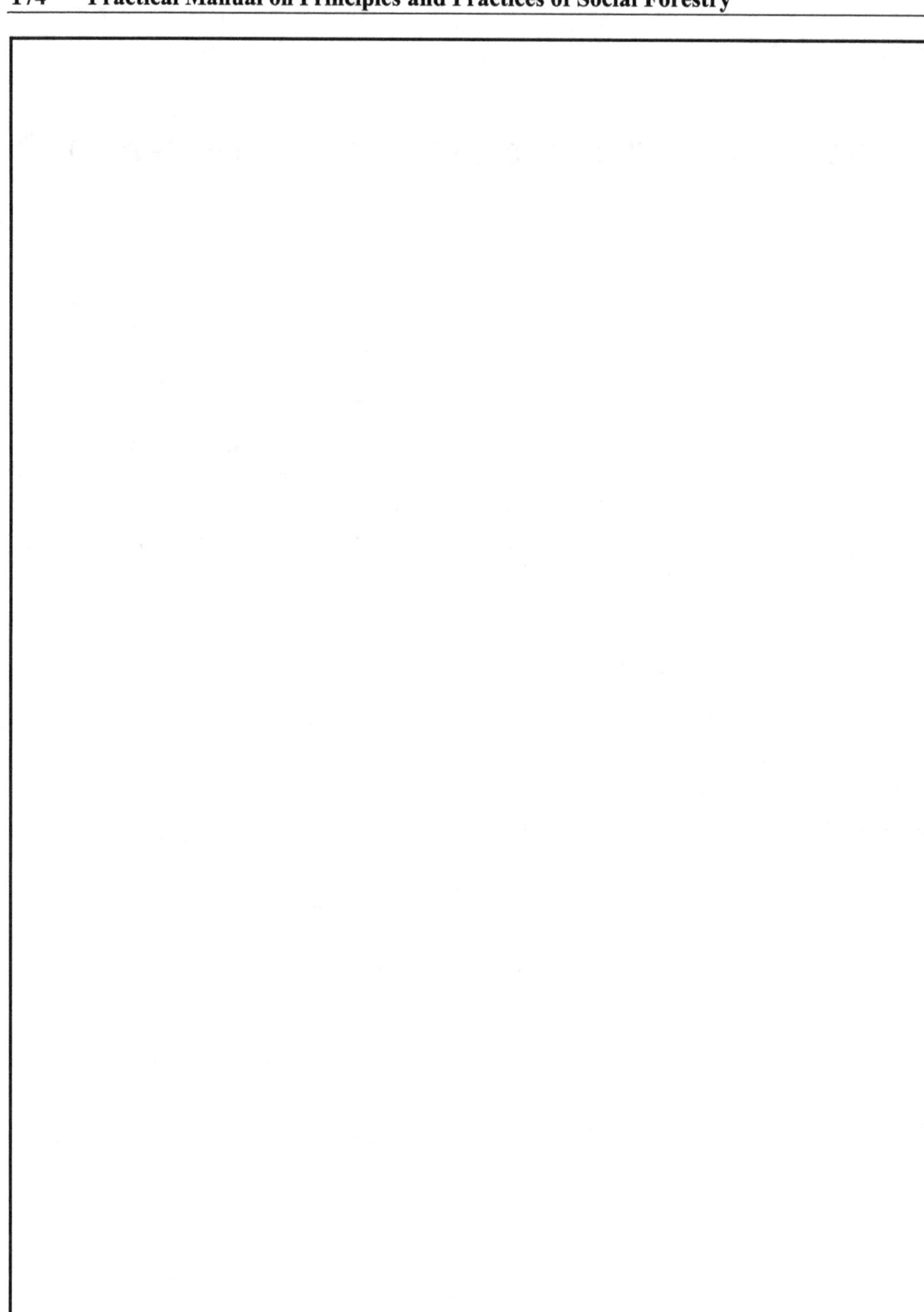

Properly designed shelterbelts shown the effect over a distance of 40 times the height of the windbreak. 1. A triangular quite zone starting from top of the windbreak and extending near to the ground level of a distance of 8 h (i.e. 8 times the height of the windbreak), where the turbulence velocity fluctuations of wind are reduced and below the values of approach flow. 2. Above and downwind of the quiet zone is a wake zone where the turbulent fluctuations are greater than those in approach flow. The degree of protection offered by a windbreak depends upon the orientation, height, density, species composition and spacing of the windbreak.

1. *Orientation*: Windbreaks should be oriented at right angles to the prevailing or problem winds, because the protected zone has maximum extension in the down wind direction.

2. *Height*: The protected zone associated with windbreak is directly proportional to the height of windbreak. Hence the height should be from the top of the crop canopy and should be at least 2-3 times the height of the crop.

3. *Density*: Moderately dense windbreaks significantly reduce the wind velocities (50-60%). Windbreaks with lower densities are generally in effective for most crop protection purposes. The moderately dense windbreaks do not cause, as much downwind turbulence as dense windbreaks, hence they are more effective.

4. *Species composition*: The most desirable plants for a variety of field windbreaks are those that have a density of about 50-80% in single row planting. Their spread not exceeding 3 meters and should have a potential height of 5-30 meters. Tall, narrow crowned species usually provided the greatest benefit for the amount of land taken out of production. Casuarina, Eucalyptus, Sesbania, *Parkinsonia aculeate* are highly suitable windbreaks.

5. *Spacing*: In general within the row spacing are 1 to 2.5 meters of shrubs, 1.5-2.5 meters for small trees, 2-6 meters for medium to tall trees. Normally the spacing between the rows varies from 2-6 meters.

Characteristics of tree species suitable for windbreaks

1. Resistance to drought.
2. Resistance to frost and wind throw. i.e. should be deep rooted.
3. Tolerance to temperature extremes.
4. Easy to establish and long lived.
5. Fast growing with tall uniform shape.
6. Resistance to pests and diseases.
7. Should not serve as alternate host for fungi, insects and other pests.

Advantage of windbreaks and shelterbelts

1. The shelterbelts exercise their influence on agricultural crops by preventing soil erosion and by modifying microclimate of the fields. Changes in air temperature and soil temperature and moderate.
2. They reduce wind velocity and consequent movement of the sand and soil is arrested.
3. At higher wind velocities, the stomatal apparatus tends to be decreased, the belts create more favourable conditions for the plant growth in the sheltered area.
4. They slow down the wind and conserve precipitation.

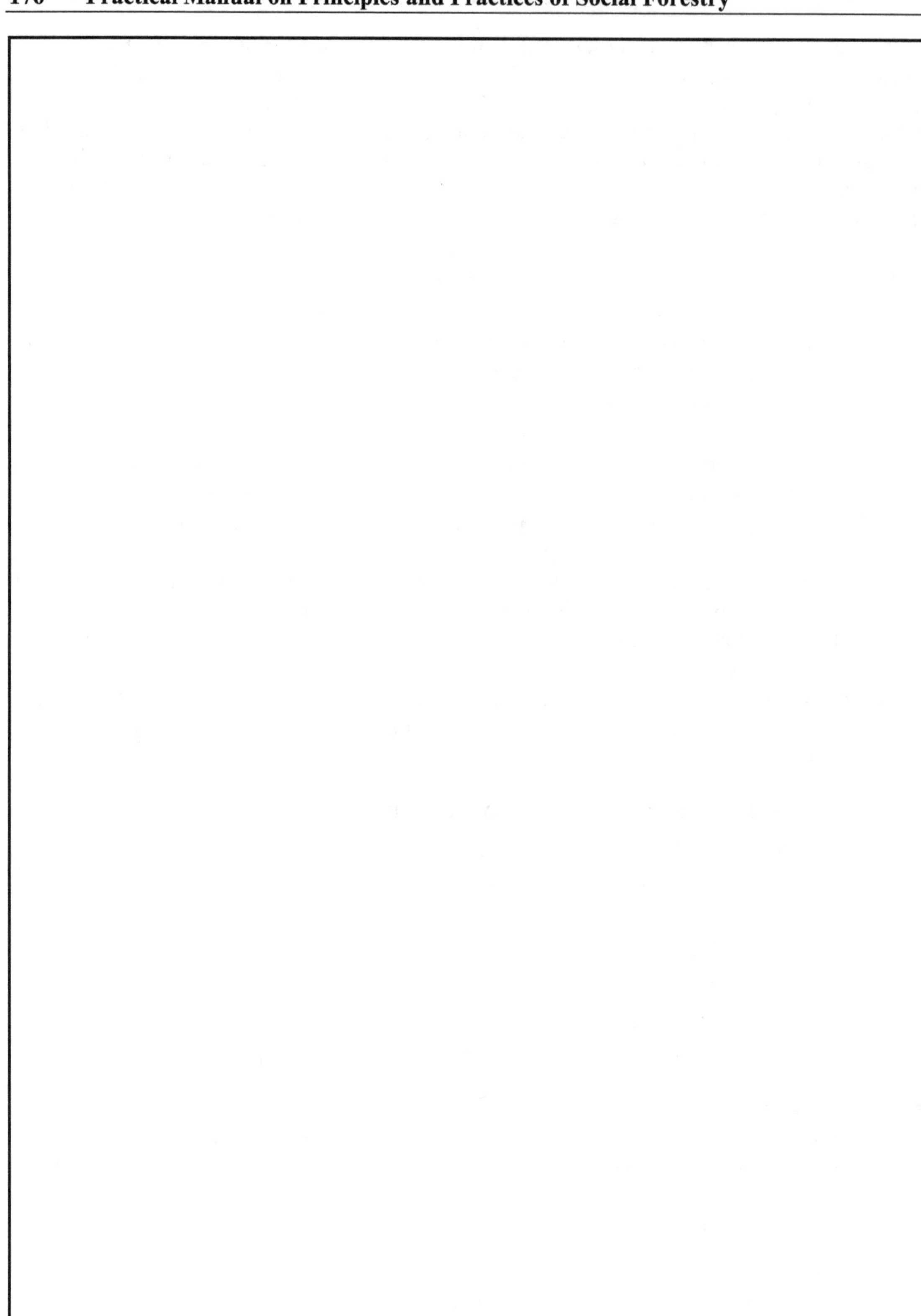

5. Dewfall in sheltered areas has been found to be 200% greater than on exposed ground.

6. The increase in soil moisture content on the protected areas varied from 0.3 to 7.8 percent.

7. Shelterbelts also protect the cattle from the ill effects of hot and cold winds.

8. Shelterbelts can protect orchards as they reduce wind damage to flowers and fruits.

9. Shelterbelts supply fuel and also increase the bird population, which kills the pests.

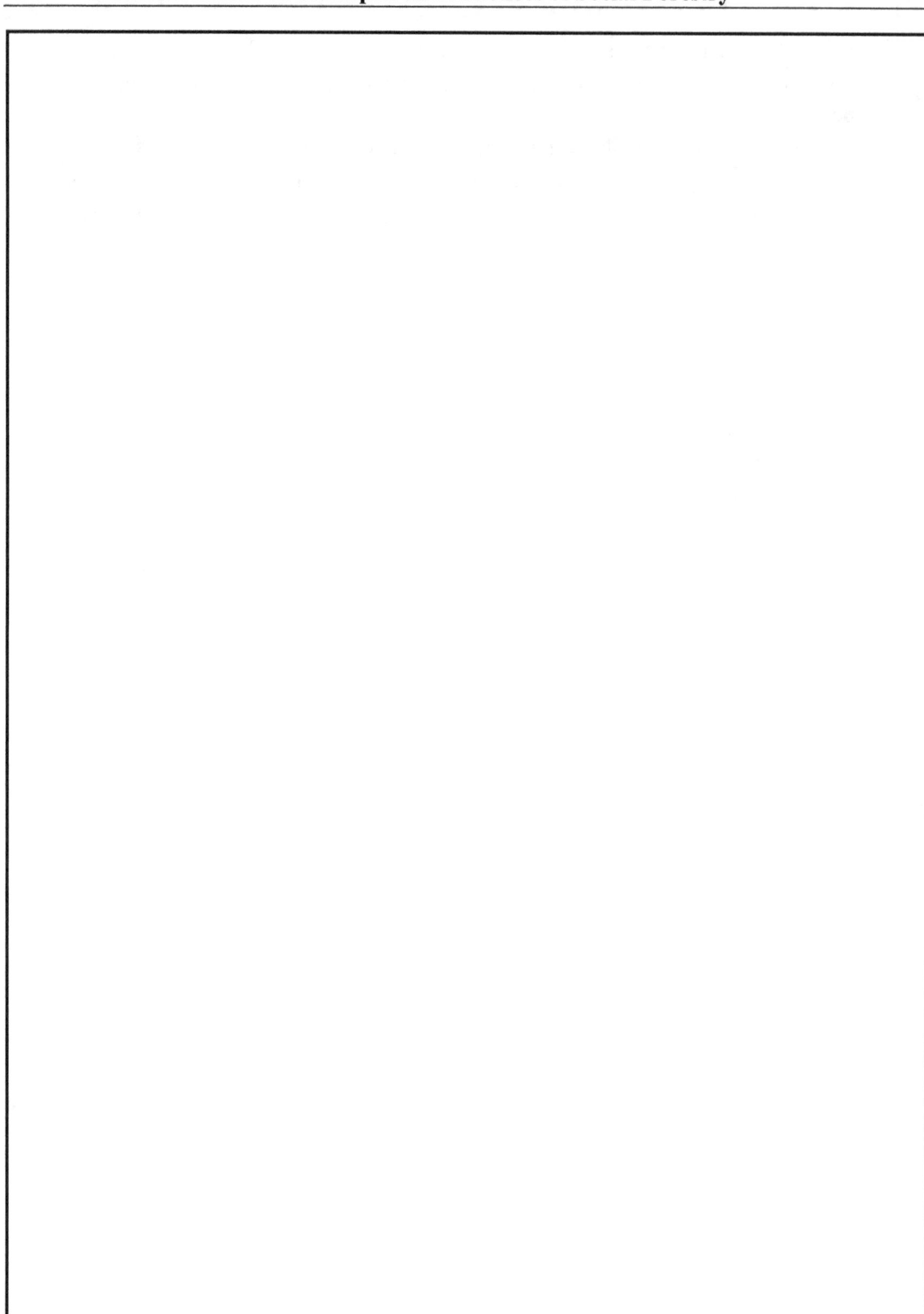

MAJOR FOREST PRODUCTS

Objectives:

1. ____

2. ____

3. ____

TIMBER

1. Teak (*Tectona grandis*)

A.   Physically recorded dry weight 15-20 kgs/cu.ft. small characteristic and powerful like old shoe-leather, very offensive when being worked.

Grains coarse and open but much smooth wood in fast grown timber, surface greasy to touch, dull.

Bark grey ½″ thick fibrous with shallow longitudinal wrinkles peeling off in long thin flakes.

Colour uniform dark golden – yellow, brown, grayish brown, sapwood white or dirty white.

B.   **Anatomic characters (*Transverse section*)**

   (a)   *Pores*: Very prominent 8-20 per sq.mm. usually single in one or two rows.

   (b)   *Rays*: Just visible, medium, uniform 4-6 per mm.

   (c)   *Rings*: Very clear on account of dense pore-ring and looser springwood following the denser tissue of Autumn.

2. Rose wood (*Dalbergia latifolia*)

A.   *Physical characters*: Recorded dry weight 20-30 kgs per cu.ft smell fragrant like rose water.

Grain : Very even, though moderately coarse and open, surface bright, the pores dull the rays shining, the ground bright.

Bark: Grey or grayish brown ½ -3/8″ thick, irregular, short   cracks, exfoliating in thin fibrous longitudinal flakes.

B.   **Anatomical characters**

   (a)   *Pores*: Inconspicuous thought clear, no contrast, evenly distributed, rare 0-6/mm.

   (b)   *Rays*: Very obscure 6-8/mm.

   (c)   *Rings* : Obscure, yet well defined. The bands of colour and narrow black pigment zones not necessary indicating the annual growth.

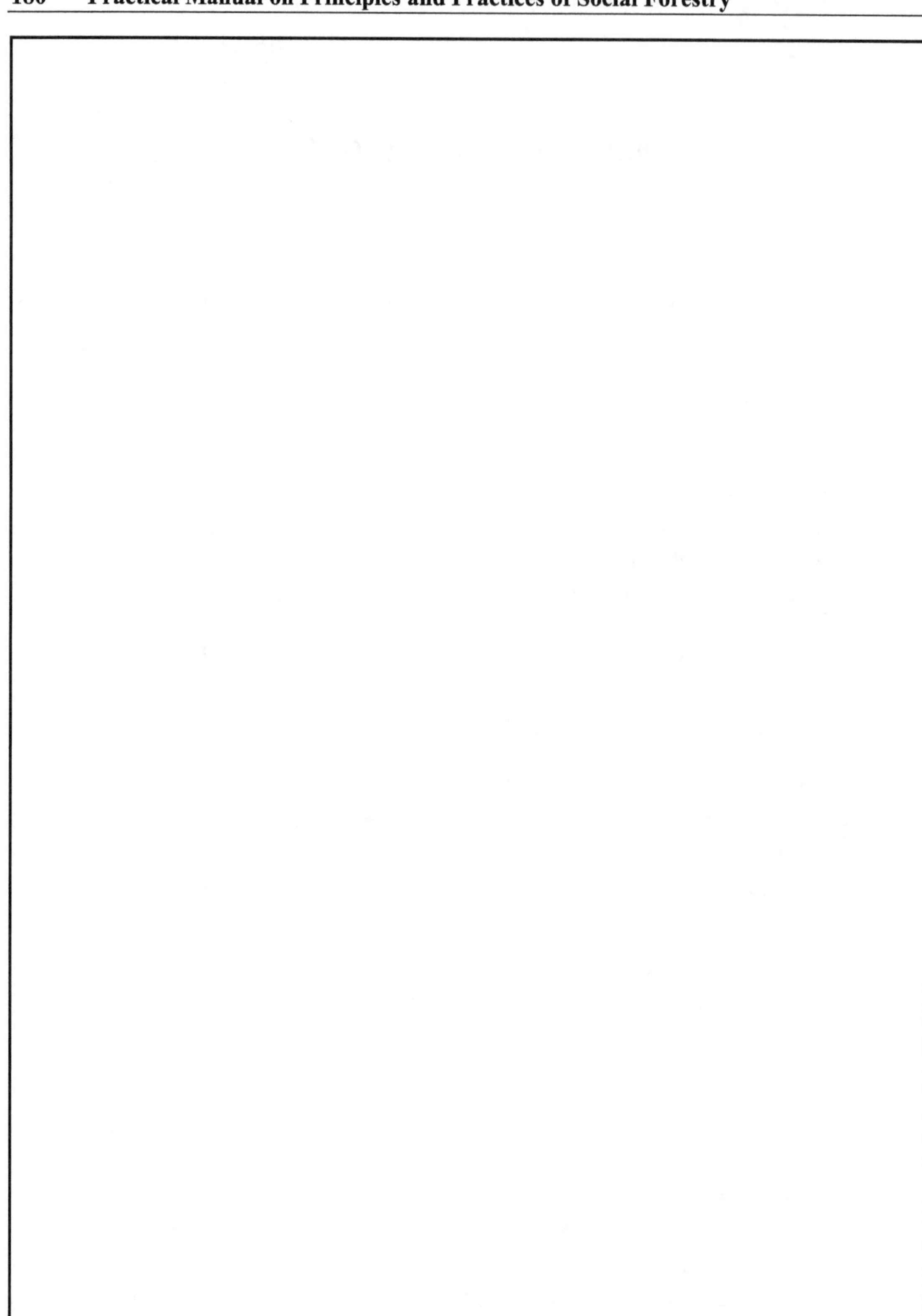

3. Red sander *(Pterocarpus santalinus)*

Bark	:	Blackish brown, deeply cleft with vertically and horizontally into rectangular plants.
Wood	:	Extremely hard, sapwood white, heartwood dark claret-red to almost black, but always with a deep red twinge, orange red when freshly cut.
Pores	:	Moderate sized, scattered, very scanty, joined by fine pale undulating concentric lines at undulating concentric lines at un-equal distances, difficult to see on an old specimen.
Rays	:	Fine, numerous, equidistant.

4. Sal *(Shorea robusta)*

Bark	:	1-2" thick, dark coloured, rough, with deep longitudinal furrows.
Wood	:	Sap wood small, whitish heartwood brown, or reddish brown darkening on exposure. Wood hard to very hard 25 kgs/ cu.ft. Annual rings indistinct. Invisible only in young trees.
Pores	:	Moderate sized to large often filled with resin.
Rays	:	Uniform moderately broad, straight, very preeminent joined by short white transverse lines.

5. Bijasal *(Pterocarpus marsupium)*

Bark	:	1/3 inch thick, grey, with long vertical cracks exfoliating in small pieces.
Wood	:	Very hard, close grained, giving a red resin, sapwood small, heartwood yellowish brown, with darker streaks.
Pores	:	Moderate sized and large, often sub-divided scanty, resinous, uniformly distributed in pale patches.
Rays	:	Very fine, numerous, short, uniform and equidistant.

6. Axle wood *(Bakli) Anogeissus latifolia*

Bark	:	Smooth, whitish gray
Wood	:	Grey, hard shining, smooth with a small purplish brown, irregularly shaped very hard heartwood annual rings marked by liens without pores.
Pores	:	Small, very numerous.
Rays	:	Very fine, extremely numerous, uniform, equidistant

7. Gamari *(Gmelina arborea)*

Bark	:	¼ inch, smooth white or whitish grey.
Wood	:	Yellowish, grayish or reddish white, glory luster, even grained, soft, light but strong, durable, does not warp or crack. Annual rings marked either by a white line or by more numerous pores in the spring wood.
Pores	:	Large and moderate sized often sub-divided.
Rays	:	Short, moderately broad, prominent, visible in silver grain as irregular horizontal bands.

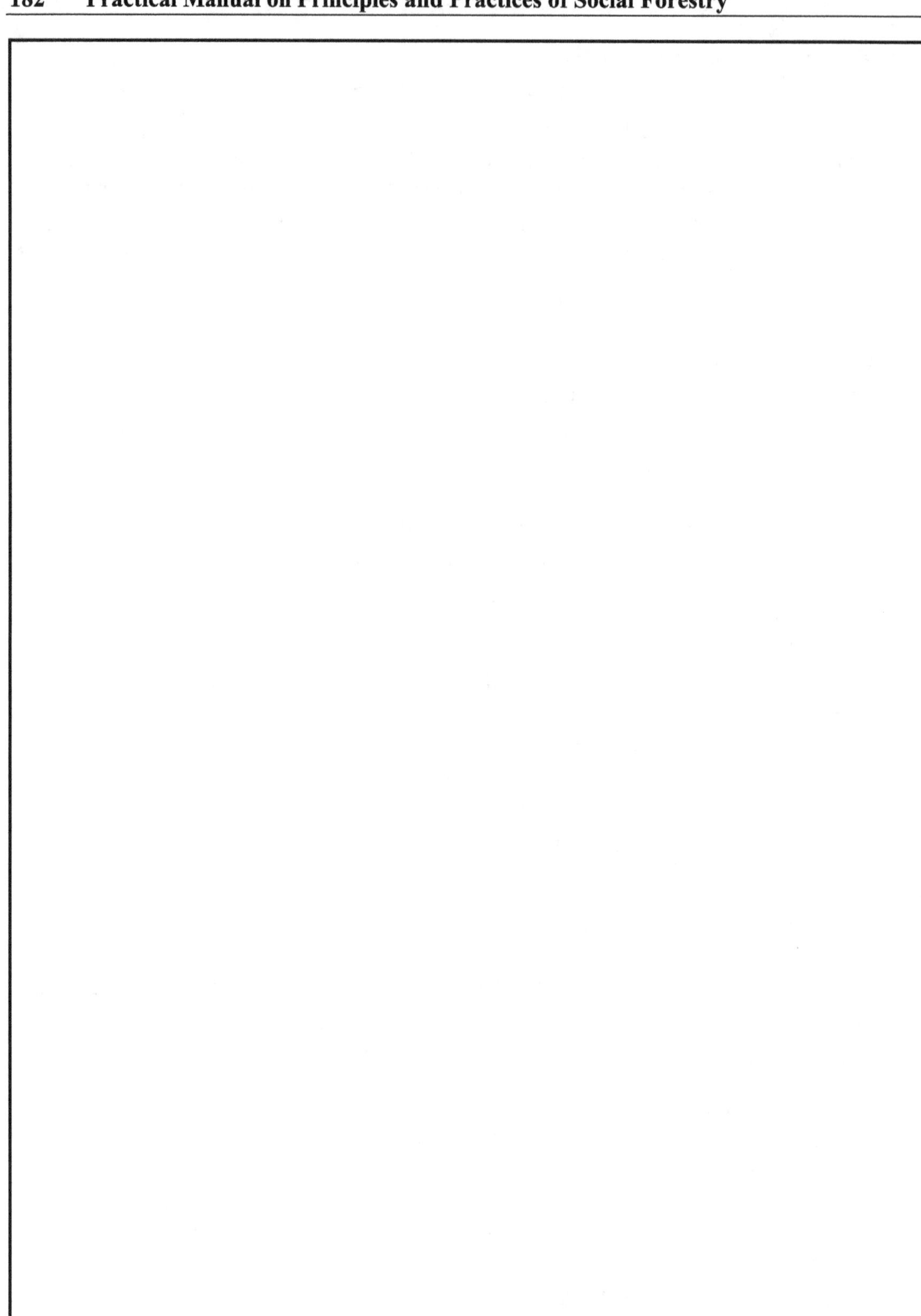

8. **Nallamaddi (*Terminalia tomentosa*)**

Bark	:	1" thick, grey to black with long broad, deep longitudinal fissures and short transverse cracks.
Wood	:	Sapwood reddish white, heartwood dark brown, hard beautifully variegated with streakes of darker colour.
Pores	:	Moderate sized and large, uniformly distributed.
Rays	:	Not district, very fine, numerous uniform, equi-distant often wavy.

9. **Anduku (*Boswellia serrata*) (*Salai*)**

Bark	:	½" thick, yellow or greenish yellow, exfloriating in small hard irregular flakes or thin plates.
Wood	:	Moderately hard, smooth, sapwood white, heartwood brown.
Pores	:	Scanty, moderate sized often subdivide containing resin.
Rays	:	Moderately broad, very short, not very numerous.

10. **Sissoo (*Dalbergia sissoo*)**

Sapwood distinct from heartwood, pale yellowish-white heartwood golden brown to deep brown with darker streaks wood hard, heavy wood structure is similar to rose wood.

11. **Semul (*Bombax ceiba*)**

Bark	:	Grey when young with conical prickles with corky base, when old with long, irregular vertical cracks.
Wood	:	White when fresh cut, turning dark on exposure, very soft, perishable, no heartwood, no annual rings.
Pores	:	Very scanty, very large.
Rays	:	Fine to broad, numerous, not prominent. Pores and silver grain prominent on vertical section.

12. **Mulberry (*Morus alba*)**

Bark	:	Brown, Rather rough.
Wood	:	Hard, sapwood white, Heartwood yellow or yellowish brown, darkening on exposure. Annual rings marked by conspicuous belt of moderate sized and large pores.
Pores	:	Scanty
Rays	:	Fine to moderately broad. Rather numerous, giving a pretty silver grain.

CLASSIFICATION OF MINOR FOREST PRODUCTS

I. Fibres and Flosses

(a) Fibres from leaves e.g : Screw pines, Sago palm, Avage or sisal fibre, plantain fibre, wild date and palmyra palms, golpatha.

(b) Stem and Beast fibre : *e.g: Streculia villosa, Hardwickia binata, Bauhinia vallic, Grewia oppositifolia, Ficus religiosa, Ficus bengalensis, Hemp (Cannabis sativa) Calatropis gigantea.*

(c) Flosses from *Bombax ceiba, Cochlospermum gossypium, Calotropis* spp.

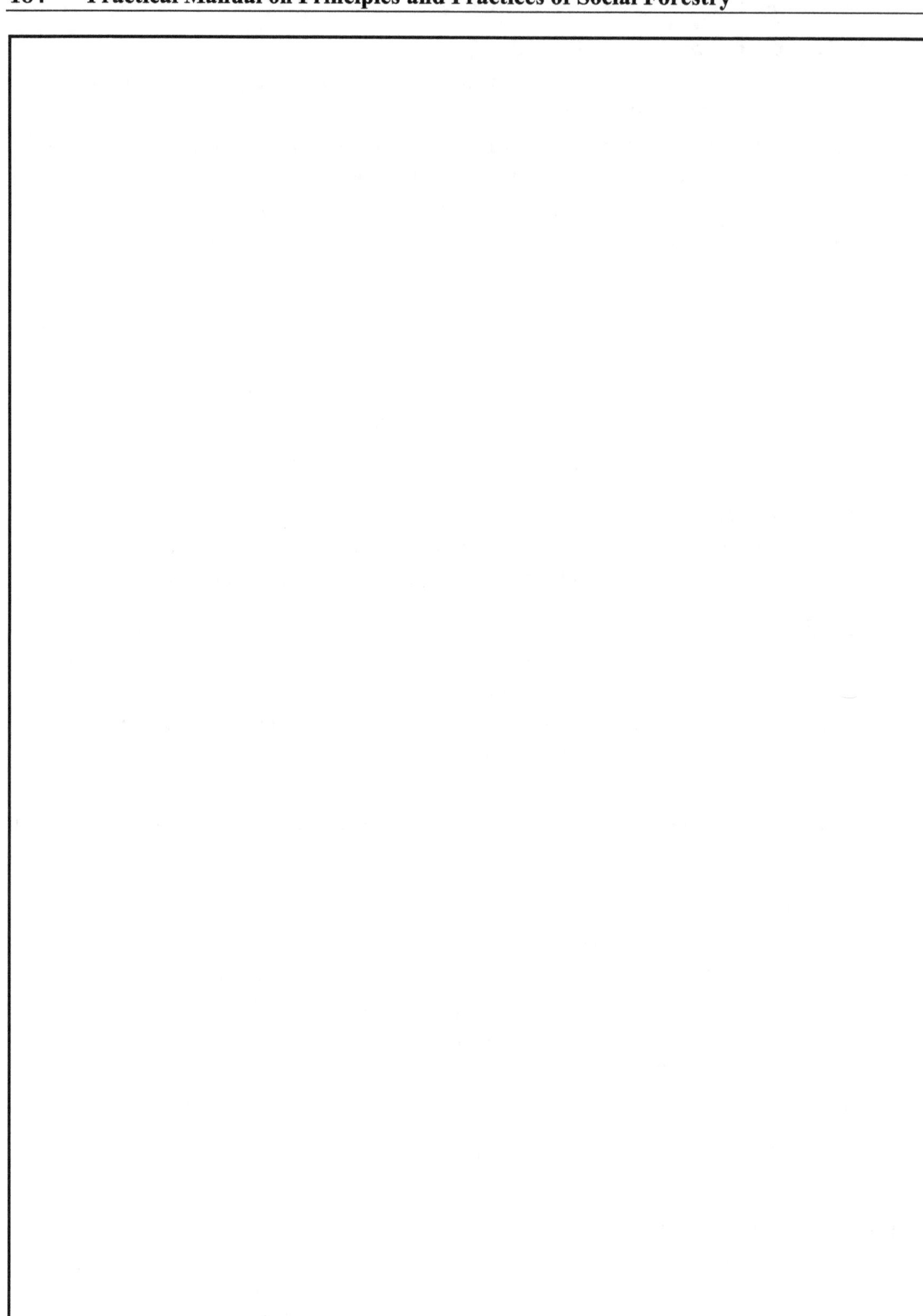

II.   Grasses (other than oil grasses)

(a)   Fibre grasses, Munj grass, *(Saccharum arundinaceae)* bhabar or sabaigrass *(Ischoemum angustifoluim)* useful for paper making, mats, cordage, chicks, basket making etc.

(b)   Thatching grasses : *Andropogon contortus* (spreading grass), *Imperata arundinacea.*

(c)   Grazing and fodder grasses :

(d)   Other fibrous products :

    (i)   Khus – Khus *Vetiveria zizanioides* (roots of commerce)

    (ii)   Korai *(Cyperus tegetum)* for mat making

    (iii)   Reeds sur reeds *(Pharagmites spp.)*

III.   Distillation products

(a)   Grass Oils

    (i)   Rosha grass oil *(Cymbopogon martini)*

    (ii)   Lemongrass oil *(Cymbopogon flexuosus)*

(b)   Wood Oils :

    (i)   Sandal wood oil *(Santalum album)*

    (ii)   Agar – agar oil from Agar or Eagle tree *(Aquilaria agallocha)*

    (iii)   Other wood oils from *Cedrus deodara*

(c)   Miscellaneous products of distillation :

    (i)   Cutch and katha from *Acacia catechu*

    (ii)   Charcoal

    (iii)   Camphor from *Cinnamum camphora*

    (iv)   Cinnamon oil from *Cinnamum zeylanicum*

    (v)   Mohwa liquor from *Madhuca latifolia*

IV.   Oil Seeds: From neem, karanj, mohwa, etc.

V.   Tans and Dyes

(a) Tan barks	:	From *Acacia arabica* bark, *Cassia fistula* bark Mangrova bark *(Rhizophora spp.)*
(b) Tan leaves	:	Leaves of *Anogeissus latifolia, Emblica officinalis, Lawsonia alba.*
(c) Tan Fruits	:	From fruits of *Terminalia chebula* (Hirda or Harra tree) Babul pods, are also used.
(d) Wood Bark and Root Dyes	:	*Pteracarpus santalinus, Artocarpus integrifolia* (Wood dye). *Punica granatum (Pomegranate)* for root dyes. *Terminalia tomentosa* for bark dyes.
(e) Dye flowers	:	*Eg*: Cedrella toona flowers. *Butea monosperma.*

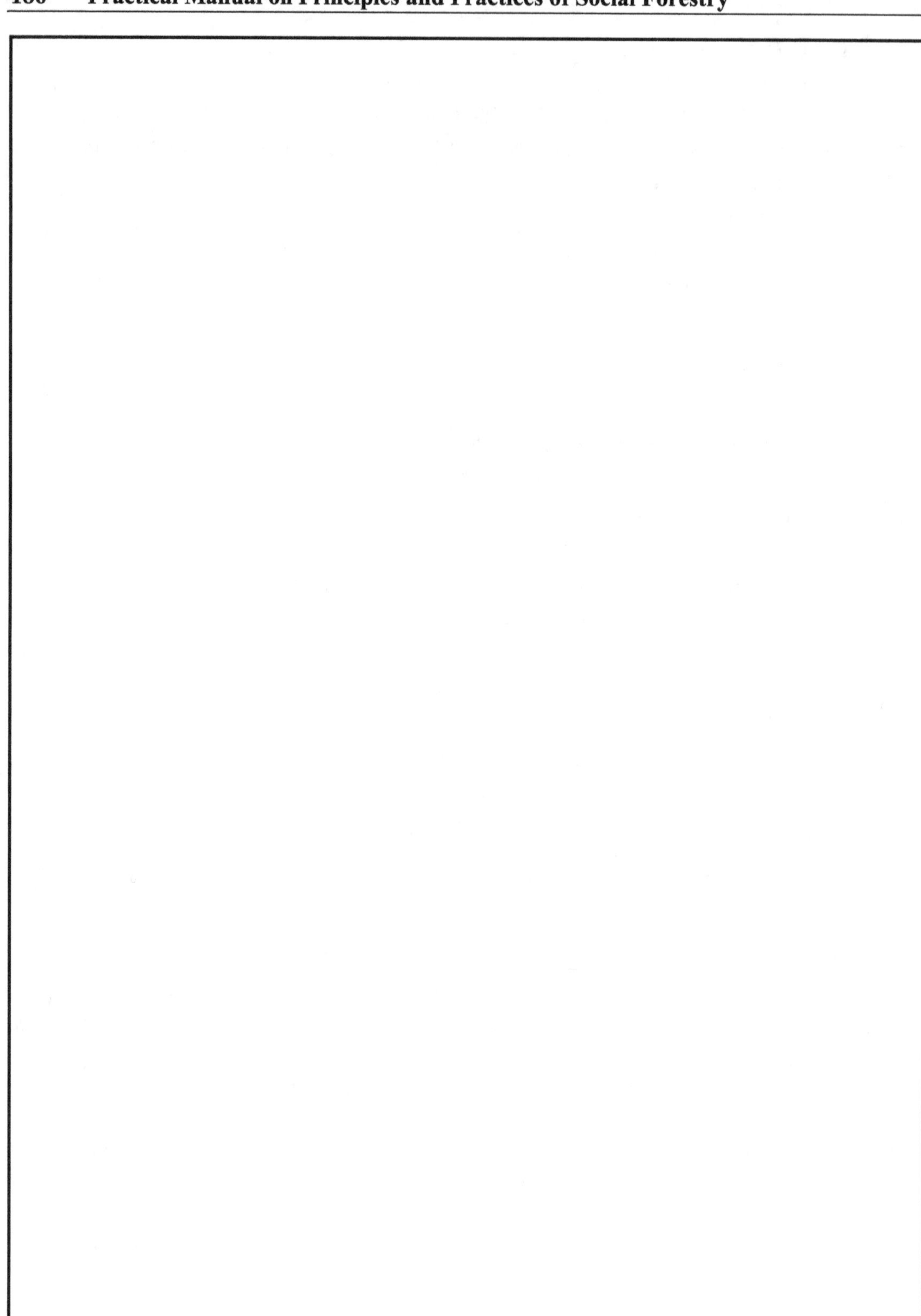

VI. Gums, Resins and oleoresins:

(a) Gums and gum resigns

: Gum Arabic from *Acacia Senegal* Babul gum from Babul tree. Gurkino from *Pterocarpus marsupium*.

(b) Resins other than pine resin

: From *Callophyllum inophyllu* and also from *Hopea* spp.

(c) Pine-resin from *Pinus roxburghii*

(d) Oleo resins from *Dipterocarpus turbinalis*

VII. Rubber from *ficus elastica*

VIII. Drugs and Spices

(a) Root drugs

: *Podophyllum emodi, Berberis aristata*

(b) Stem and bark drugs

: From *Cinnamum zeylanicum, Ailanthus excelsa* bark. *Ficus religosa*

(c) Fruit and seed drugs and spices

: Nux-vomica seed, ritha or soapnut, pepper, shikaki, bael fruit, *Cassia fistula* pods, marking nut.

IX.   Edible products

: Jamun, ber, pala, apricot, mango, jackfruit.

X.   Bamboos and Canes

: *Dendrocalmus strictus, Bambusa arundinacea (Bamboos) Calamus Spp – Canes.*

XI.   Animal products :

(a) *Lac.*

(b) *Wax and Honey.*

XII   Miscellaneous Minor Forest Products : Beedi leaves. *(Diospyros melanoxylon)* or Abrius or turkey.

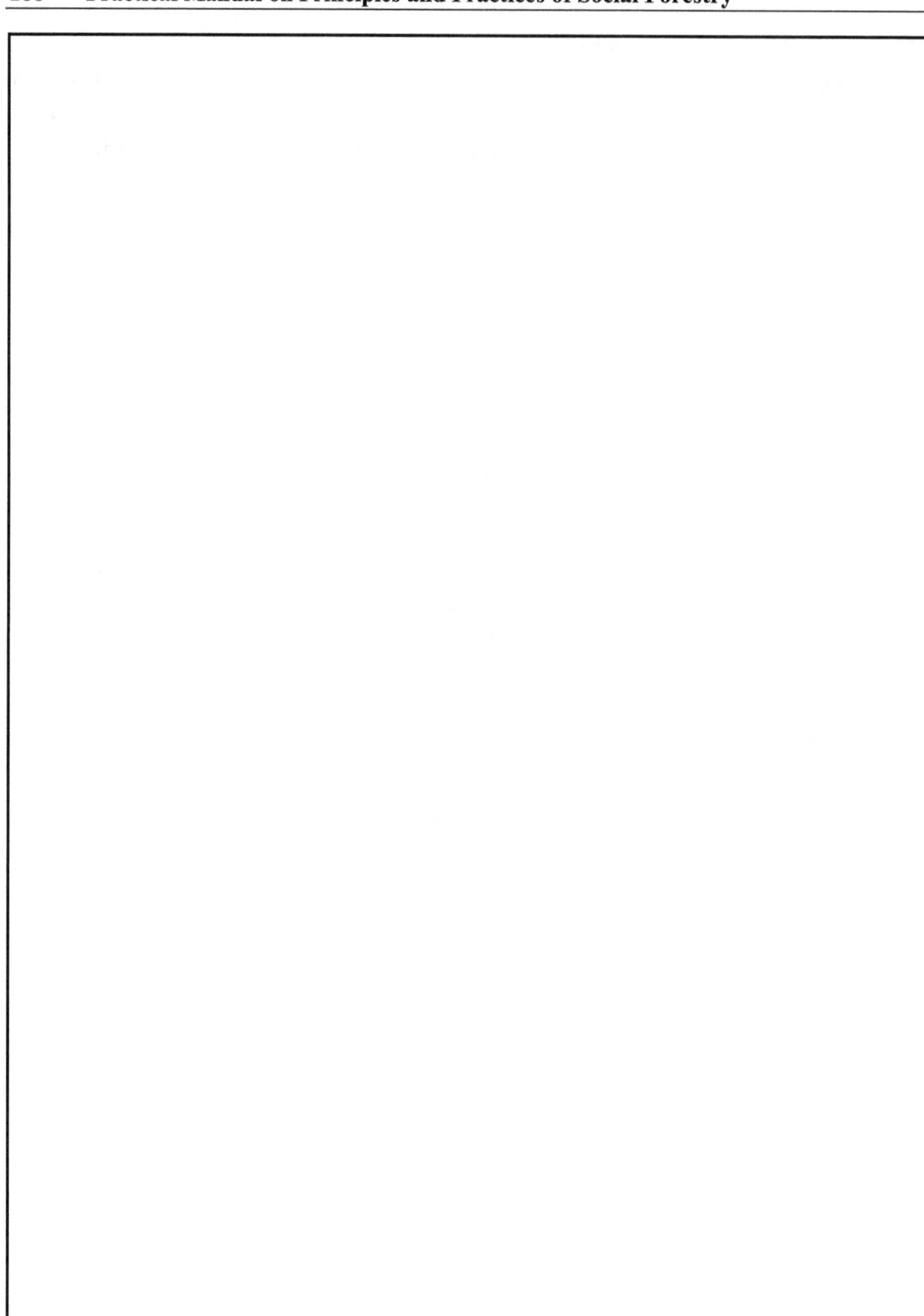

TREE TRANSLOCATION

Translocation of trees is very important as the survival rate of the trees is more than ninety percent. During translocation, one should be careful not to damage the roots and cover them with enough soil. The ball of earth around the roots should not be distuirbned and the tree is to to be translocated after pruning the branches .All the leaves have to be cut to avoid further transpiration and the tree is lifted with proclainer along with the ball of earth and it should be shifted to a new place where it is to be planted. It takes a month for the tree to recover its new growth.

As part of the redevelopment of the Ginkgo trees have been lifted from the concrete rings in which they were planted in during the early 1990's.

The trees are root pruned to remove any broken or damaged roots and wrapped in hessian with a wire retaining basket to keep them warm and to avoid drying out.

The trees are transported in a trailer to Acton Park and pulled by a specialist vehicle to minimise any damage and compaction to the grassed area next to the lake where the trees are to be planted.

The tree pits are excavated using a mini digger with tracks reducing the amount of soil compaction around the tree pit. Trees are then lifted and guided into the pits using a sling from the mini digger attached to the tree lifting equipment. Once placed upright the tree is anchored using a purpose designed underground anchoring system that is driven into the ground.

Things then start to take shape and then the finished product.

Any time when you come across people mercilessly cutting trees at any place, stop them and call Andhra Pradesh Real Estate Developer's Association (APREDA), which helps in tree translocation and save trees. The Association will provide necessary assistance for translocation of the trees.

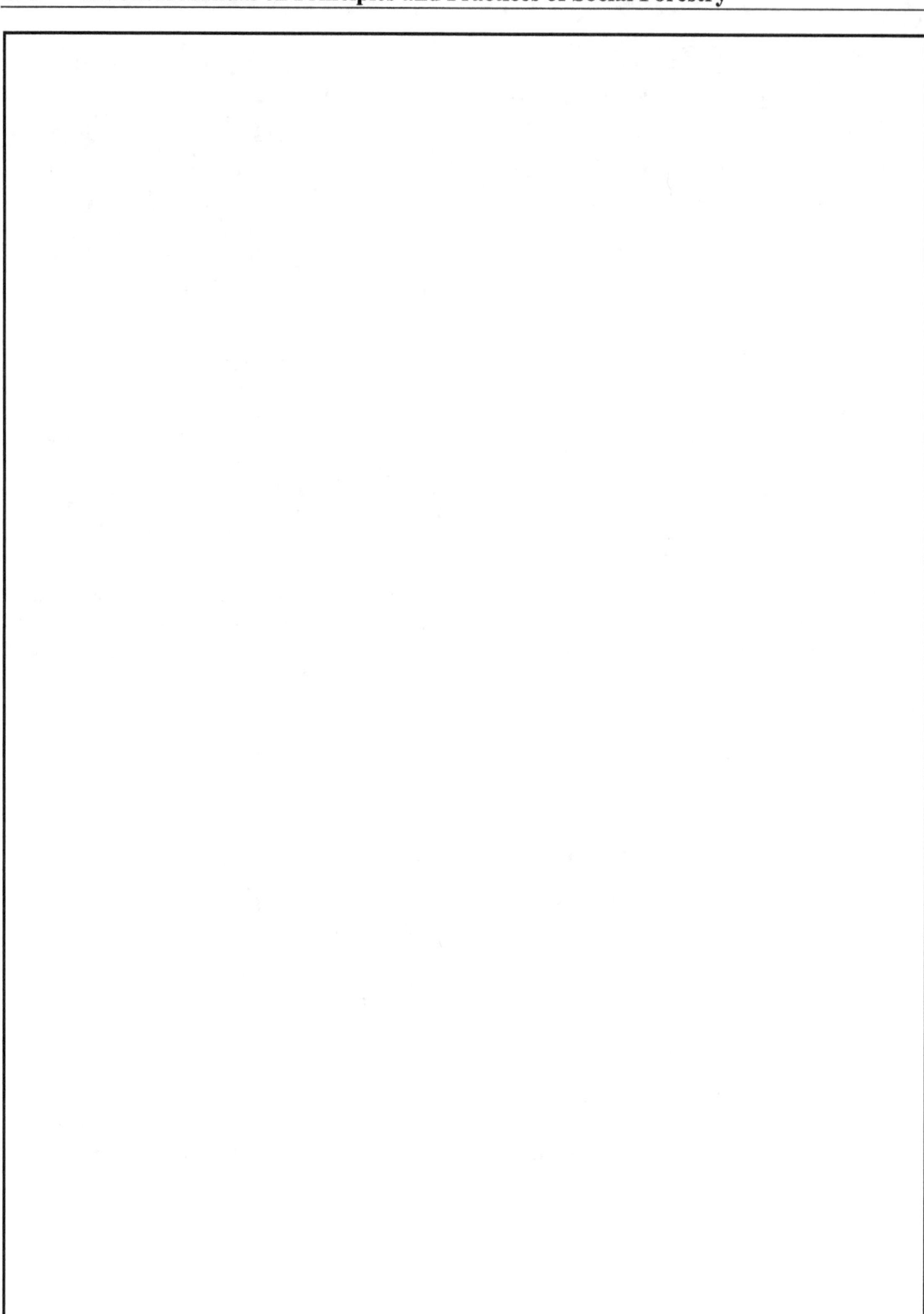

COST OF CULTIVATION OF COMMERCIAL TREES IN WASTELANDS: BAMBOO

Objectives:

1. ____

2. ____

3. ____

Harvesting and Yield: The annual yield of bamboo clump depends on the number of new culms produced each year. This in turn is related to the production of young culms. Culms mature after 2 to 3 years. To maximize shoot output, some shoots must be left each year to develop into leafy young culm. It is reported that bamboo clump on an average produce 10 culms in a year under good growing conditions. Considering a 30 year life cycle, one clump may produce 300 culms on the whole.

The harvesting can be done from 5th year onwards, however for commercial production, harvesting will start from 6th year. In the first year of harvest, i.e. after 6 year of planting, 6 culms /clump will be harvested followed by 7 culms /year, in 7th year, 8 culms year on 8th year onwards and the yield stabilizer at 10 culms/year after 8th year onwards. The culms which are of 1 or 2 year old are generally left for regeneration. Considering the average weight of each culm at 10 kg, the yield in 1st year of harvest is 9.6 tonnes/acre, which will stabilize at 14.4 tonnes by 9th year.

Production of bamboo is only the starting point. The real benefits accrue from value added products handicrafts such as mats, baskets, tools, toys and utensils and furniture are established possibilities. There are emerging industrial and large scale applications too in the manufacture of wood substitutes and composite, energy, charcoal and activated carbon. Building and structural components represent vast possibility for enterprise, value addition income and employment.

The role of bamboo in community agroforestry as a means of generating income to the rural poor is very important.

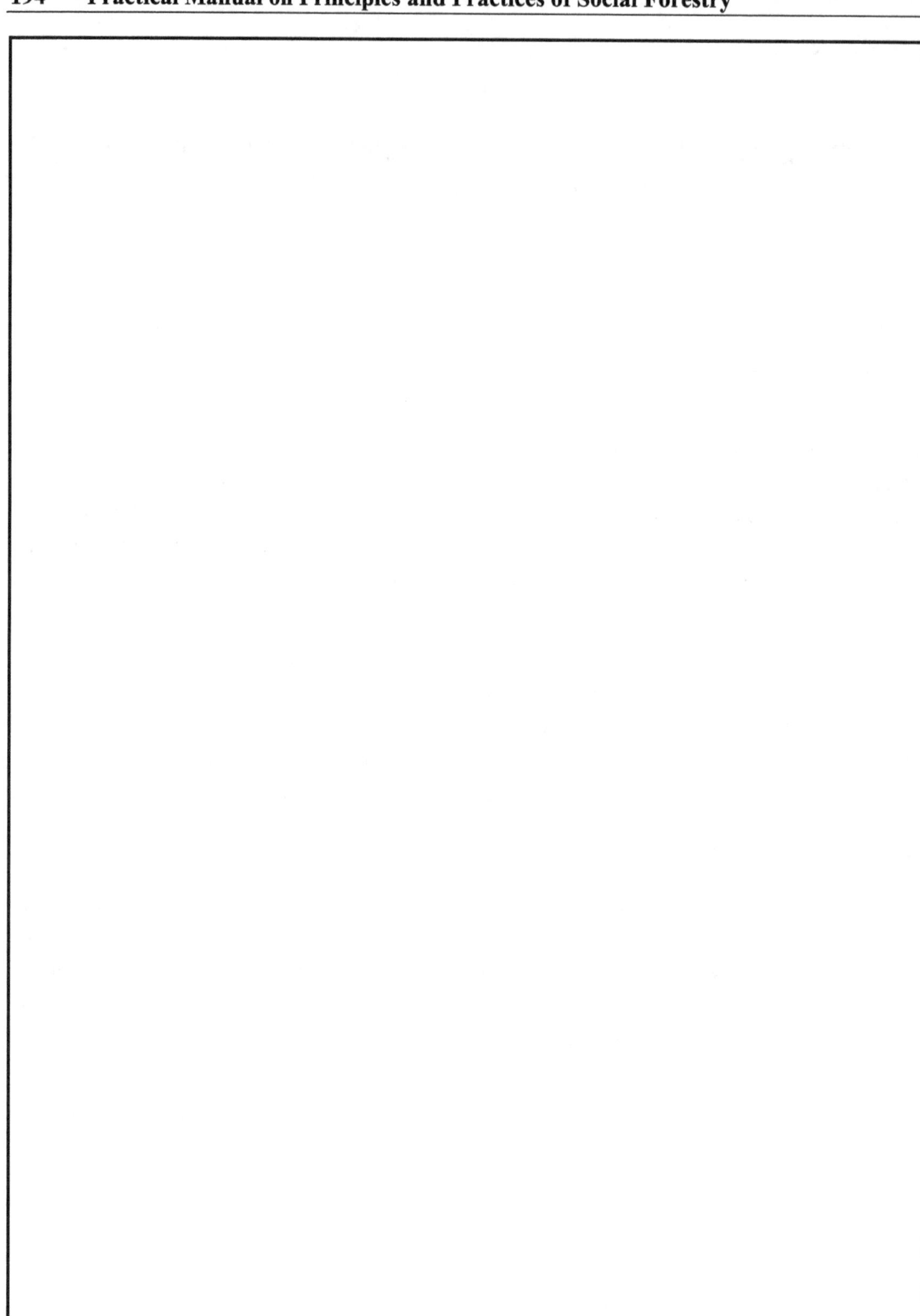

COST OF CULTIVATION OF COMMERCIAL TREE CROPS

Bamboo cost of cultivation on 1 acre wasteland.

S. No	Description of items	Units	I year	II year	III year	IV year	V year	Total
	Inputs cost							
1	Planting material including 20% morality replacement in II year.	@ Rs 5/- each seedlings	200 nos @ 5/- 1000	50 nos @ Rs 5/- 250	-	-	-	1250
2	Manure and fertilizers	1 acre unit	500	500	500	350	350	2200
3	Plant protection		200	200	200	200	200	1000
4	Irrigation 3 times /year	@ Rs 400/- irrigation	1200	1200	-	-	-	2400
5	Fencing (biofence)	Rs 20/- running meter	6400	-	-	-	-	6400
	Sub total		9300	2150	700	550	550	13250
	Cost of labour	Man days						
1	Land preparation	5 days @ 250/day	1250	-	-	-	-	1250
2	Pit digging, refilling of pits 15 pits/MD	200 pits	3750	1250	-	-	-	5000
3	Planting and staking	5	1250	500	-	-	-	1750
4	Plant protection	3	300	300	300	300	300	1500
5	Weeding (thrice in first year & twice in II year-4 MD /weeding	5	1250	500	-	-	-	1750
6	Pruning from III year onwards	5	0	0	500	500	500	1500
7	Soil working and others	2	500	500	500	500	500	2500
8	Harvesting -10 MD in 6th&7th year 12 MD in 8 th years onwards	10	-	-	-	-	-	-
	Sub total		8300	3050	1300	1300	1300	15250
	Contingency	5%	400	400	400	400	400	2000
	Grand total		18000	5600	2400	2250	2250	30500

Estimated Income

Year	Culms/clump	Total culms available	Income @ Rs 20/culm
5th	4	352	7040
6th	6	528	10560
7th	8	704	14080
8th year onwards	10	880	17600

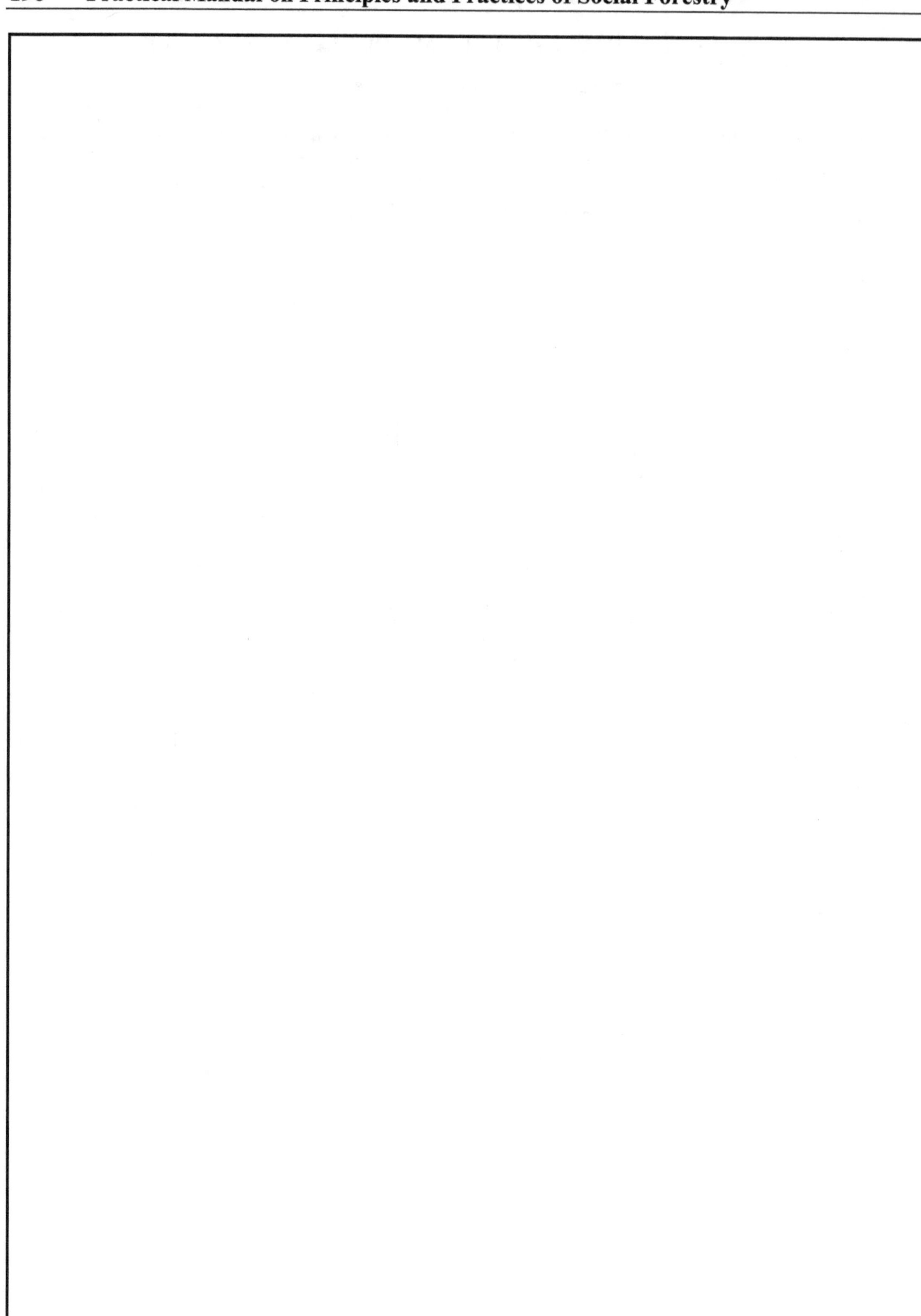

CALCULATIONS

1. Mr. Krishna Reddy from Kanakamavidi village of Moinabad mandal is having 60 acres of wastelands. He is interested to take up Bamboo plantations with a spacing of (5m × 5m). Find out how many number of Bamboo seedlings he has to purchase from the nursery including 10 per cent of seedlings for mortality replacement.

2. Mr. Sudhakar Reddy from Ibrahimpatnam mandal interested to grow 20,000 seedlings Subabul seedlings in his field. He wanted to plant the seedlings with a spacing of (2m × 2m). Find out how much area of the land in acres, he has to allocate to plant the 20,0000 seedlings of subuabul in his field. If the cost of seedlings is Rs 5/- each, find out how much amount of money the farmer has to spent to procure 20000 seedlings including 10 per cent mortality replacement.

3. Mr.Krishna Reddy from Appareddygudem village of Chevella mandal is having 20 acres of class III land. He is interested to take up the Jamun plantations in his field by planting 20,000 seedlings. Find out what is the spacing he has to adopt between row to row and plant to plant while planting the seedlings in his field.

4. How much area is needed for raising the nursery of 6 kg *Acacia nilotica* seed with plant spacing of (20 cm × 5 cm), plant percent – 50, number of seeds per kg – 1000.

5. Calculate the quantity (weight) of seedlings required for sowing an area of 300 sq. meters with a spacing of 20cm × 5cm and 1000 seeds per kg.

SOLUTIONS

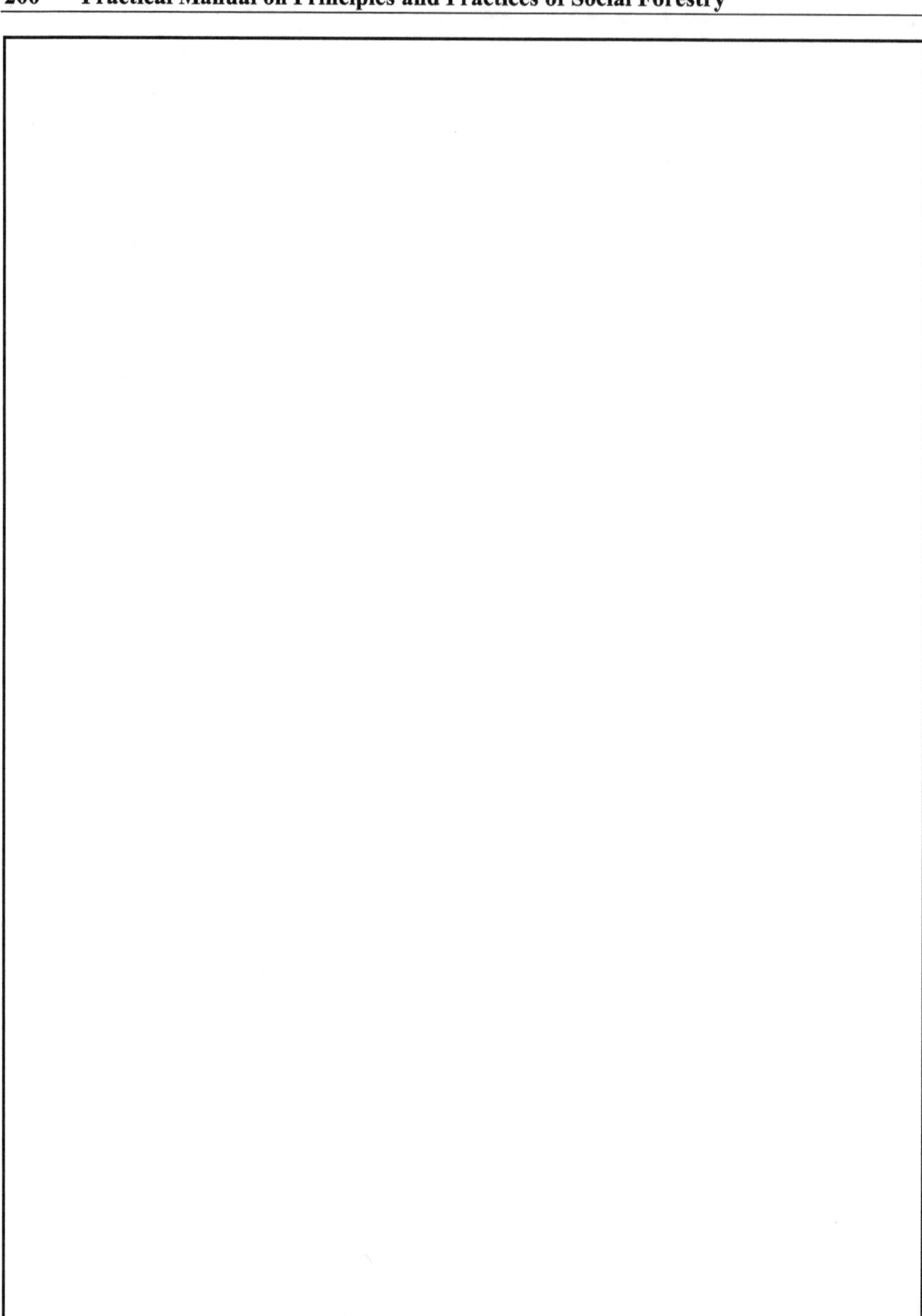

Agricultural Units of Measurements Used for land in India

1 acre	=	4840 sq. yard
	=	4046.8 sq. meters
	=	43560 sq. feet
	=	0.4047 hectare
1 hectare	=	2.4711 acres $\approx$ 2.5 acres
	=	10,000 sq meters
12 inches	=	1 foot
1 inch	=	2.54 centimeters
36 inches	=	1 yard
3 feet	=	1 yard
9 square feet	=	1 square yard
121 square yards	=	1 gunta
40 guntas	=	1 acre
1000 millimeters	=	1 meter

Glossary

A

Afforestation

Establishment of plantations on treeless land.

Alternate

Terms used of leaves or buds of broad-leaved trees that arise first on one side of a twig, then on the other.

Annual ring

The sleeve of wood put on each growing season over the previous year's wood and under the bark on both stem and branches. Consists of <u>spring wood</u> and summer wood, the latter usually darker and thereby showing up as a ring in cross section.

Arboretum

A collection of specimen trees.

Arboriculture

Management of individual trees or groups of trees primarily for their amenity value.

Assortment (Stand)

The breakdown of a stand of timber into different products. Estimated using stand assortment tables, and based on the average <u>dbh</u> of the stand and the minimum top diameter required for the products being considered.

B

Bark

Protective layer on the outside of stems and branches, consisting of living cork cells on the inside and dead cells on the outside.

Bars

See <u>Pallet wood</u>.

Beating up

Replacing trees that have died shortly after planting.

Billets

Pieces of small diameter round timber cut to length.

Blaze

The mark made by an axe slicing off bark from a tree, normally done to indicate the tree is to be felled. Also the act of marking.

Blueing stumps

See Dyeing stumps.

Bole

Main part of the stem of a tree before it separates into branches.

Brashing

Removal of the lower dead branches, up to about two metres, of trees in a stand.

Breast height

1.3 metres (43 inches) above the ground on the highest side. Point at which diameter or girth is measured on a standing tree.

Bud scale

Scale that covers and protects a developing leaf or flower.

Burr

Excrescence on base of tree. Some broad-leaved trees with a burr can be very valuable – much sought after by craftspeople and carpenters.

Butt

Lower part of the stem of a tree.

Butt rot

The most serious, common place disease in British forestry. Fungal infection of the roots and lower stem of the living tree degrading the most valuable part of the tree. *Heterobasidion annosum* (formerly *Fomes annosus*) is the most important cause of butt rot. See also Dyeing stumps.

C

Canopy

The foliage and small branches of tall trees in a wood when these have interlaced to form continuous cover.

Catkin

Male or female flowers hanging in chains: they lack coloured petals because they are wind-pollinated flowers.

Clone

Identical series of plants arising from a single parent by artificial or natural vegetative propagation.

Compartment

A unit within the forest, demarcated (for administrative purposes) by permanent features e.g. roads and streams.

Compound

Term describing leaf that consists of several leaflets, e.g. ash leaf.

Conifer

Tree on which the seeds are borne in a cone.

Coppicing

Cutting of woody stem at ground level to encourage growth of several stems from one root system.

Crown

Branches and upper part of the stem of tree.

Cultivar

Variations of a species arising in cultivation and propagated for some unusual characteristic, such as leaf colour or shape.

Current annual increment (C.A.I.)

Volume increment of a stand in one year, or averaged over a short period of years, measured in cubic metres per hectare.

D

Deciduous tree

One that sheds all its leaves in winter.

Diameter at breast height (dbh)

The standard way to measure standing trees using a <u>girth tape</u> and measuring at 1.3 metres above ground level.

Die-back

Condition whereby individual twigs and branches die, starting from the tip. Can be one of several causes.

Dyeing Stumps

Covering the freshly cut surface of a conifer stump with urea solution which prevents colonisation by a wood rotting fungus which could otherwise spread from infected stumps to nearby standing trees. The solution is dyed to facilitate application. See also <u>Butt rot</u>.

E

Ecosystem

All plant and animal life living in a particular habitat and, to at least some extent, dependent upon each other.

Epicormics

Twigs sprouting directly out of the main stem.

Establishment

The first five to ten years or formative period that ends once young trees are of sufficient size that, given adequate protection, they are likely to survive at the required stocking. The stage in the growth of a young plantation when it no longer needs <u>beating up</u> or weeding.

Exotic

Tree introduced from overseas.

F

Felling

Cutting down trees, particularly mature trees.

Fire break

An unplanted strip left between plantations, or on their margins, which is kept clear of flammable vegetation in the early years.

Flushing

The bursting of buds in spring.

Forest

A large area dominated by trees, both conifers and broad-leaved, either planted or natural. Usually taken to include a complex landscape comprising of woodland, open space, water and settlements. See also <u>wood or woodland</u>.

Forest floor

The comparatively clear ground under a closed-canopy forest.

Forestry

Management of forests and woods.

G

Germination

Development of seedling from fertilised seed.

Glaucous

Term describing waxy film on the surface of a leaf or stem, giving it a bluish appearance and serving to reduce water loss.

Grafting

Artificial union of aerial parts of one plant with the vigorous rootstock of another.

H

Habitat

Situation in which a particular plant or animal lives.

Hardwood

The wood of broadleaved trees, a term sometimes used for the broadleaved trees themselves.

Heartwood

Dead wood consisting of several annual rings at centre of tree trunk or branch, no longer water-conducting tissue but providing structural support.

Heavy thinning

Removal of a large volume per unit area.

Hectare

Unit of land area equal to 10 000 square metres. There are 100 hectares in a square kilometre.

Hybrid

A natural or artificial cross between two species, or occasionally between two genera, when it is an 'inter-specific hybrid'.

Hypsometer

An instrument used from ground level for measuring the heights of trees.

I

Increment

The amount of new wood put on by a tree or a stand in a year or in the period between thinning measured either in cubic metres or in cubic metres per hectare.

K

Knot (live/dead)

A live knot is the base of a living branch that is incorporated in the timber. A dead knot is the remains of a dead branch that has been left in the timber.

L

Layering

Term describing the development of a new individual plant from a branch or stem that has rooted into the ground.

Leading shoot

Main shoot that develops from the terminal bud at the top of a tree each year. Also called the 'leader'.

Lenticel

Small pore in bark or a leaf for breathing.

Light thinning

Removal of a small volume per unit area.

Line thinning

A method of mechanical thinning, as distinct from selective thinning, in which a whole line of trees is taken out irrespective of their quality. Usually confined to first thinning.

Lop & top

Woody debris from cutting operation.

Lopping

Cutting branches off a tree.

M

Maiden

A tree that has grown from a seedling (naturally or planted) and which has not been coppiced.

Mean annual increment (M.A.l.)

The average rate of volume increment of a stand from time of planting to the present day, measured in cubic metres per hectare.

Mensuration

The science of measuring used in forestry to mean the measurement of standing and felled timber.

Monoculture

Growing one species as a crop.

Mounding

Type of pre-planting cultivation whereby a weed free raised planting position is created. Often used on wet sites. See <u>Dolloping</u>.

Multi-purpose forestry

Forest management that delivers multiple benefits eg. social, economic, environmental.

N

Native species

Species that have arrived and inhabited an area naturally, without deliberate assistance by man.

Natural regeneration

Young seedlings that have arisen from seed falling from trees nearby, either as a direct response to specific forest management, or by natural seeding. Very often just referred to as 'regeneration' but technically this includes artificial regeneration as well, i.e. planting.

Naturalised

An introduced tree or plant that now regenerates naturally and is widespread.

New planting

Establishing woodland on ground that was not woodland in the recent past.

Nurse

A tree species in a mixture that is planted to protect, or enhance the growth of, a more sensitive species intended to form the final crop, e.g. pine nursing oak or larch nursing beech. The nurse crop is usually removed during the early thinning stage.

O

Opposite

Term used of buds and leaves of broad-leaved trees that are arranged opposite each other in pairs on the twig.

Overbark

The volume of wood including the bark. Can be either standing volume or felled volume.

Overstorey

Trees forming the upper canopy of a forest.

P

p.year (planting year)

The year a crop was planted, e.g. p 83 = planted in 1983. ('p' should always be written in lower case).

Palmate

Leaves that have lobes arranged like the fingers of a hand, e.g. horse chestnut.

Pedunculate

Describing fruits, which are borne on a stalk (a peduncle).

Pinnate

Term used of leaf completely subdivided into several leaflets ranged along either side of midrib.

Pole length

Tree felled and debranched.

Pole stage

A plantation in the early stages of thinning.

Pollarding

Lopping of the branches of a tree at head height or a little above to encourage shoots to arise above the reach of browsing animals.

Propagation

Regeneration of new plants by means other than seed, e.g. by rooting cuttings.

Pruning (usually referred to as high pruning)

Removal of branches usually on selected stems only, above the height of Brashing with the object of reducing the knots in the timber being formed.

Pulpwood, chipwood and woodwool

Small diameter material that is destined to be fragmented before manufacture into paper pulp, building board or packing material respectively.

R

Regeneration

The production of a new crop by artificial or natural regeneration. See <u>Natural regeneration</u>.

Restocking

The replanting of an area after trees are removed.

Ring barking

Cutting round the bark and destroying the cells carrying the tree's food. Also called Girdling.

Root stock

Roots on which desirable grafts are grown.

S

Sapwood

Living wood making up the outer annual rings of tree trunk through which water from soil is conducted up the tree.

Scarification

The breaking-up of the ground surface by machine prior to planting. Used on well-drained ground.

Seedling

Young tree grown from seed, either in the forest naturally or in a nursery, prior to being transplanted or planted out in the forest.

Sessile

Describing fruits which are not borne on a stalk.

Shade bearer

A tree that will grow in the shade of others.

Shelter belt

Trees and shrubs planted in a comparatively narrow strip to provide protection, usually of farmland.

Shrub layer

Shrubs and bushes, i.e. woody plants without a single main stem, occurring well below the forest canopy.

Silviculture

Cultivation of trees as crop with the primary objective of producing wood products.

Softwood

The wood of coniferous trees or conifers themselves.

Spring wood

Thin-walled cells laid down in stems and branches in the early part of the growing season. It is of lighter colour than the summer wood in most species.

Stand

A fairly uniform collection of trees, composed of one or a few species, comprising one age class, from either artificial or natural regeneration.

Sucker

Shoot arising directly from a root or at base of a stem.

Summer wood

Thick-walled cells laid down in the middle of the growing season in stems and branches. Usually darker than the adjoining spring wood and thus showing up as a ring in cross-section.

Sustainable development

Development that meets the needs of the present without compromising the ability of future generations to meet their own needs.

T

Tap root

Main, downward-growing root.

Thinning

Periodic removal of trees in a stand which are competing with those better trees which are expected to form the final crop. The object is to benefit the final crop trees, and to get income from the thinnings before they die.

OR

A proportion of stems removed in order to give the best stems space and light to grow into a more valuable crop. This is usually carried out sometime after canopy closure and may be repeated at intervals. It is a necessary operation in the production of quality timber. A temporary reduction in standing volume will result.

Transplant

Young tree that has been grown from seed and transplanted within the nursery in preparation for being planted out in the forest.

U

Underbark

The volume of wood excluding the bark.

Undercut

Severing the taproots of seedling trees to promote growth of fibrous roots.

Underplanting

Planting a new crop below an existing crop with the intention that it will eventually succeed the older crop.

Understorey

Trees and/or shrubs below the canopy. See also <u>Overstorey</u>.

V

Variety

Variation of a species arising in the wild, usually differing on only one characteristic, such as leaf colour or leaf shape.

Vegetative propagation

Reproduction by cuttings, layering or grafting and not involving fertilisation.

W

Weeding

Removing unwanted herbaceous growth. See also <u>cleaning</u> and <u>spot weeding</u>.

Whip

A thin tree with very small crown reaching into the upper canopy. Also used for thin whippy broad-leaved tree as compared with a 'standard' or a forest transplant.

Whorl

Structures such as buds and leaves arising three or more at a time around a stem at the same point.

Wood or woodland

A smaller area of trees than a forest, usually applies to broad-leaved trees. The distinction between a "forest" and a "wood" or "woodland" is simply one of scale, and to some extent the words are interchangeable.

Wood raw material equivalent

The volume of trees required to produce a wood product. Can be measured underbark or overbark.

Y

Yield class

A classification based on height growth, used to assess the volume production of a stand. Class values range from four (cubic metres per hectare per annum) for broadleaves, larch and pine, up to 30 for grand fir. It reflects the potential productivity of the site for the tree species growing on it.